METHODOLOGY FOR THE DIGITAL CALIBRATION
OF ANALOG CIRCUITS AND SYSTEMS

METHODOLOGY FOR THE DIGITAL CALIBRATION OF ANALOG CIRCUITS AND SYSTEMS

with Case Studies

by

Marc Pastre

*Ecole Polytechnique Fédérale de Lausanne,
Switzerland*

and

Maher Kayal

*Ecole Polytechnique Fédérale de Lausanne,
Switzerland*

A C.I.P. Catalogue record for this book is available from the Library of Congress.

ISBN-10 1-4020-4252-3 (HB)
ISBN-13 978-1-4020-4252-2 (HB)
ISBN-10 1-4020-4253-1 (e-book)
ISBN-13 978-1-4020-4253-9 (e-book)

Published by Springer,
P.O. Box 17, 3300 AA Dordrecht, The Netherlands.

www.springeronline.com

Printed on acid-free paper

All Rights Reserved
© 2006 Springer
No part of this work may be reproduced, stored in a retrieval system, or transmitted
in any form or by any means, electronic, mechanical, photocopying, microfilming, recording
or otherwise, without written permission from the Publisher, with the exception
of any material supplied specifically for the purpose of being entered
and executed on a computer system, for exclusive use by the purchaser of the work.

Printed in the Netherlands.

Contents

List of Figures		xi
List of Tables		xvii

1.	INTRODUCTION		1
	1	Context	1
	2	Objectives	2
	3	Compensation methodology	2
	4	Applications of the compensation methodology	2
	5	Book organization	3
2.	AUTOCALIBRATION AND COMPENSATION TECHNIQUES		5
	1	Introduction	5
	2	Matching	5
		2.1 Matching rules	6
		2.2 Matching parameters	6
	3	Chopper stabilization	7
		3.1 Principle	7
		3.2 Analysis	8
		3.3 Implementation	9
	4	Autozero	11
		4.1 Principle	11
		4.2 Analysis	12
		4.3 Noise	14
	5	Correlated double sampling	18
	6	Ping-pong	18
	7	Other techniques	20

	8	Classification	21
	9	Conclusion	22
3.	DIGITAL COMPENSATION CIRCUITS AND SUB-BINARY DIGITAL-TO-ANALOG CONVERTERS		23
	1	Introduction	23
	2	Digital compensation	23
	3	Successive approximations	24
		3.1 Principle	25
		3.2 Working condition	28
		3.3 Reverse successive approximations algorithm	29
		3.4 Complexity	31
	4	Sub-binary radix DACs	31
		4.1 Use of sub-binary DACs for successive approximations	31
		4.2 Characteristics	32
		4.3 Resolution	34
		4.4 Tolerance to radix variations	34
	5	Component arrays	35
		5.1 Sizing	36
	6	Current sources	38
		6.1 Current-mirror DAC	39
	7	R/2R ladders	40
	8	Linear current division using MOS transistors	41
		8.1 Principle	41
		8.2 Second-order effects	45
		8.3 Parallel configuration	45
		8.4 Series configuration	46
	9	M/2M ladders	48
		9.1 Principle	48
		9.2 Complementary ladder	49
		9.3 Second-order effects	50
		9.4 Trimming	51
	10	R/xR ladders	51
		10.1 Principle	51
		10.2 Working condition	53
		10.3 Terminator calculation	54
		10.4 Terminator implementation	55
		10.5 Ladder sizing	57
		10.6 Terminator sizing	58

	10.7 Radix	60
11	$M/2^+M$ ladders	62
	11.1 M/3M ladders	62
	11.2 M/2.5M ladders	64
	11.3 Ladder selection and other $M/2^+M$ ladders	65
	11.4 Current collector design	67
	11.5 Complementary ladders	72
	11.6 Layout	72
	11.7 Measurements	73
12	Comparison	77
13	Linear DACs based on $M/2^+M$ converters	78
	13.1 Principle	78
	13.2 Calibration algorithm	81
	13.3 Radix conversion algorithm	84
	13.4 Digital circuit implementation	85
	13.5 Analog circuit implementation	87
	13.6 Compensation of temperature variations	90
	13.7 Comparison with other self-calibrated converters	90
14	Conclusion	91

4. METHODOLOGY FOR CURRENT-MODE DIGITAL COMPENSATION OF ANALOG CIRCUITS   93

1	Introduction	93
2	Two-stage Miller operational amplifier	93
3	Compensation current technique	96
	3.1 Detection configuration	97
	3.2 Detection node	100
	3.3 Compensation node	105
	3.4 DAC resolution	113
	3.5 Low-pass decision filtering	114
	3.6 Continuous-time compensation	115
	3.7 Up/down DAC	117
4	Simulation with digital compensation circuits	124
	4.1 Principle	125
	4.2 Automatic compensation component	126
	4.3 Compensation component during adjustment	128
	4.4 Compensation component during compensation	130
	4.5 Multiple digital compensation	133
	4.6 Example of implementation for PSpice	134

		4.7	Offset compensation of the Miller amplifier	136
	5	Application to SOI 1T DRAM calibration		138
		5.1	1-transistor SOI memory cell	139
		5.2	Memory cell imperfections	140
		5.3	Sensing scheme	141
		5.4	Calibration principle	144
		5.5	Calibration algorithm	146
		5.6	Measurements	147
	6	Conclusion		148

5. HALL MICROSYSTEM WITH CONTINUOUS DIGITAL GAIN CALIBRATION — 151

	1	Introduction		151
	2	Integrated Hall sensors		151
		2.1	Hall effect	152
		2.2	Hall sensors	153
		2.3	Hall sensor models	155
	3	Spinning current technique		157
	4	Sensitivity calibration of Hall sensors		160
		4.1	Sensitivity drift of Hall sensors	161
		4.2	Integrated reference coils	162
		4.3	Sensitivity calibration	163
		4.4	State of the art	166
	5	Hall sensor microsystems		171
		5.1	Analog front-ends for current measurement	171
	6	Continuous digital gain calibration technique		173
		6.1	Principle	173
		6.2	Combined modulation scheme	175
		6.3	Demodulation schemes	176
		6.4	Gain compensation	179
		6.5	Offset compensation	183
		6.6	Noise filtering	184
		6.7	Delta-sigma analog-to-digital converter	189
		6.8	Rejection of signal interferences	193
	7	Conclusion		197

6. IMPLEMENTATION OF THE HALL MICROSYSTEM WITH CONTINUOUS CALIBRATION — 199

	1	Introduction	199

	2	Hall sensor array	199
	3	Preamplifier	201
		3.1 Programmable gain range preamplifier	201
		3.2 DDA	202
		3.3 Operational amplifier	207
	4	Demodulators	208
		4.1 Switched-capacitor integrators	209
		4.2 External signal demodulator	213
		4.3 Reference demodulator	216
		4.4 Offset demodulator	220
	5	Delta-sigma modulator	221
	6	System improvements	224
		6.1 Compensation of the reference demodulator offset	224
		6.2 Coil-sensor capacitive coupling	225
		6.3 External interferences	226
		6.4 Alternate modulation/demodulation schemes	227
	7	System integration	230
		7.1 Configuration and measurement possibilities	230
		7.2 Integrated circuit	231
		7.3 Measurement results	233
	8	Conclusion	240
7.		CONCLUSION	241
	1	Highlights	241
	2	Main contributions	242
	3	Perspectives	242
References			245
Index			255

List of Figures

Figure 1.	Functional chopper amplifier	7
Figure 2.	Temporal analysis of a chopper amplifier	8
Figure 3.	Frequency analysis of a chopper amplifier	8
Figure 4.	Fully differential chopper amplifier	9
Figure 5.	Implementation of a modulator/demodulator using cross-coupled switches	10
Figure 6.	CMOS transmission gate	10
Figure 7.	Demodulator for single output chopper amplifier	11
Figure 8.	Autozero amplifier principle	12
Figure 9.	Analogically compensated autozero amplifier	13
Figure 10.	Digitally compensated autozero amplifier	13
Figure 11.	Autozero baseband and foldover noise transfer functions	15
Figure 12.	Resulting noise with autozero and small amplifier bandwidth	16
Figure 13.	Resulting noise with autozero and large amplifier bandwidth	17
Figure 14.	Effect of the 1/f corner frequency on the resulting noise	18
Figure 15.	Ping-pong amplifier system	19
Figure 16.	Operational amplifier swapping	20
Figure 17.	Digital compensation of the offset of an operational amplifier	24
Figure 18.	Ideal 4-bits DAC input/output characteristics	25
Figure 19.	Equivalent offset	26

Figure 20.	Successive approximations algorithm	26
Figure 21.	Successive approximations algorithm timing	27
Figure 22.	Reverse successive approximations algorithm	30
Figure 23.	Reverse successive approximations algorithm timing	30
Figure 24.	Input/output characteristics of a radix 1.75 DAC	32
Figure 25.	Input/output characteristics of a radix 1.5 DAC	33
Figure 26.	Parallel capacitor array	36
Figure 27.	Series resistor array	36
Figure 28.	Sub-binary DAC based on current-mirrors	39
Figure 29.	Current-mode R/2R ladder	40
Figure 30.	Normalized drain current of the MOS transistor	43
Figure 31.	Current division circuit	43
Figure 32.	Current division without input current	44
Figure 33.	Current division with input	44
Figure 34.	Equivalent transistor of two transistors in parallel	46
Figure 35.	Equivalent transistor of two transistors in series	47
Figure 36.	M/2M ladder	48
Figure 37.	PMOS M/2M ladder	49
Figure 38.	Inverse M/2M ladder	50
Figure 39.	R/xR ladder	51
Figure 40.	Modified R/xR ladder	53
Figure 41.	2R terminator in a R/3R ladder	56
Figure 42.	Maximum allowable mismatch in function of xT	59
Figure 43.	Best-achievable radix with a sub-binary converter	61
Figure 44.	M/3M ladder	62
Figure 45.	M/2.5M ladder	64
Figure 46.	M/2^+M ladder selection	66
Figure 47.	Current mirror as M/3M current collector	68
Figure 48.	Voltage/current characteristics of a diode-connected transistor	69
Figure 49.	Successive approximations with current mirrors as collectors	71
Figure 50.	Layout overview of one stage of a M/2.5M converter	72
Figure 51.	M/2^+M test-chip micrograph	73

List of Figures xiii

Figure 52.	Standard deviation of the current division in M/2.5M ladders	75
Figure 53.	Standard deviation of ρ in each stage of the M/2.5M$_4$ ladder	76
Figure 54.	Standard deviation of ρ in each stage of the M/3M$_1$ ladder	77
Figure 55.	Input/output characteristics before calibration	79
Figure 56.	Input/output characteristics after calibration	80
Figure 57.	DAC system architecture	81
Figure 58.	DAC calibration principle	82
Figure 59.	DAC calibration algorithm	83
Figure 60.	DAC radix conversion algorithm	85
Figure 61.	Digital circuit implementation	86
Figure 62.	Transresistance current collector	87
Figure 63.	Regulated cascode current collector	88
Figure 64.	Single-input current comparator	89
Figure 65.	DAC micrograph	90
Figure 66.	Two-stage Miller operational amplifier	94
Figure 67.	Small-signal model of the two-stage amplifier	95
Figure 68.	Offset detection in the closed-loop configuration	98
Figure 69.	Offset detection in the open-loop configuration	100
Figure 70.	Offset measurement in the closed-loop configuration	101
Figure 71.	Offset measurement in the open-loop configuration	102
Figure 72.	Implementation of a comparator with a digital buffer	104
Figure 73.	Input/output characteristics of the CMOS inverter	104
Figure 74.	Compensation by current injection	105
Figure 75.	Offset correction by additional differential pair	107
Figure 76.	Offset correction by degenerated current mirror	107
Figure 77.	Offset correction by unilateral current injection	108
Figure 78.	Offset correction by improved unilateral current injection	110
Figure 79.	Offset correction by bilateral current injection	111
Figure 80.	Analog averaging of the offset measurement	114
Figure 81.	Digital averaging of the offset measurement	115
Figure 82.	Imperfection tracking with successive approximations	116
Figure 83.	Imperfection tracking with up/down	117

Figure 84.	Up/down current mirror principle	118
Figure 85.	Smooth transition during up/down step	119
Figure 86.	Up/down current mirror schematic	122
Figure 87.	Up/down current mirror micrograph	124
Figure 88.	2-pass simulation algorithm	125
Figure 89.	Single-ended compensation component in the schematic editor	126
Figure 90.	Differential compensation component in the schematic editor	127
Figure 91.	Single-ended compensation component netlist for the first pass	128
Figure 92.	Model of the analog feedback loop of the first pass	128
Figure 93.	Differential compensation component netlist for the first pass	130
Figure 94.	Single-ended compensation component netlist for the second pass	131
Figure 95.	Final value range of the successive approximations algorithm	132
Figure 96.	Differential compensation component netlist for the second pass	133
Figure 97.	Modified 2-pass simulation algorithm	134
Figure 98.	PSpice diode model	135
Figure 99.	Programmable current source	136
Figure 100.	Untrimmed offset of a typical Miller amplifier	137
Figure 101.	Miller amplifier offset with single-ended 8-bits trimming	138
Figure 102.	SOI 1T DRAM cell	139
Figure 103.	Read current dispersion of the 1T DRAM cell	140
Figure 104.	Retention characteristics of the 1T DRAM cell	141
Figure 105.	Reference current window as a function of time	141
Figure 106.	Sense amplifier for SOI 1T DRAM	142
Figure 107.	Sense amplifier model	143
Figure 108.	Automatic reference adjustment algorithm	145
Figure 109.	Optimized automatic reference adjustment algorithm	147
Figure 110.	Write/read cycles on 3 adjacent memory cells	148
Figure 111.	Hall effect	152

List of Figures

Figure 112.	Cross-like Hall sensor and symbol	153
Figure 113.	Cross-like Hall sensor implementation in P-substrate CMOS	154
Figure 114.	Purely resistive Hall sensor model	155
Figure 115.	Modelling of the offset of the Hall sensor	156
Figure 116.	Modelling of the offset and Hall effect	156
Figure 117.	Spinning current technique	157
Figure 118.	Sensor and preamplifier	158
Figure 119.	Typical thermal drift of the current-related sensitivity	161
Figure 120.	Integrated calibration coil	162
Figure 121.	Sensitivity calibration principle	164
Figure 122.	Influence of the calibration period on the variation of B_{ext}	166
Figure 123.	Calibration by dual signal ± reference measurement paths	167
Figure 124.	Calibration by separate signal and reference measurement paths	169
Figure 125.	Calibration by frequency separation	170
Figure 126.	System architecture	174
Figure 127.	Gain adjustment feedback loop	180
Figure 128.	Gain adjustment feedback loop with ADC and digital comparison	181
Figure 129.	Compensation current injection	182
Figure 130.	Offset correction feedback loop	183
Figure 131.	Spectral representation of the modulated reference signal	185
Figure 132.	Band-limitation of the noise to increase the SNR	186
Figure 133.	Low-pass filtering after demodulation to increase the SNR	187
Figure 134.	Demodulator and delta-sigma filter transfer functions	188
Figure 135.	Delta-sigma used as an analog-to-digital integrator	189
Figure 136.	Typical signals in the delta-sigma modulator	190
Figure 137.	Low-pass filter function of the delta-sigma ADC	193
Figure 138.	High-pass parasitic transfer function of the reference demodulator	195
Figure 139.	Parasitic transfer function before and after filtering	196

Figure 140.	Hall sensor and reference coil array	200
Figure 141.	Preamplifier block diagram	201
Figure 142.	Sensor array and first stage of the preamplifier	202
Figure 143.	Model of the DDA with 5 differential inputs	203
Figure 144.	Schematic of the DDA	205
Figure 145.	Schematic of the operational amplifier	207
Figure 146.	Switched-capacitor integrator	209
Figure 147.	Addition principle	210
Figure 148.	Subtraction principle	211
Figure 149.	Switch timing for an addition	212
Figure 150.	Switch timing for a subtraction	212
Figure 151.	External signal demodulator switch timing	214
Figure 152.	Demodulator phase shift	215
Figure 153.	Reference demodulator	217
Figure 154.	Reference signal demodulator switch timing	218
Figure 155.	Offset signal demodulator switch timing	221
Figure 156.	Delta-sigma modulator	222
Figure 157.	Delta-sigma switch timing	223
Figure 158.	Offset compensation in the gain adjustment feedback loop	224
Figure 159.	Model of the coil-sensor capacitive coupling	225
Figure 160.	Micrograph of the current measurement microsystem	232
Figure 161.	Preamplifier and demodulator output for $B_{ext} = 0$	235
Figure 162.	Preamplifier and demodulator output for negative B_{ext}	236
Figure 163.	Preamplifier and demodulator output for positive B_{ext}	237
Figure 164.	Nonlinearity measurement	238
Figure 165.	Offset drift measurement	239
Figure 166.	Sensitivity drift measurement	239

List of Tables

Table 1.	Characteristics of the compensation techniques	21
Table 2.	Successive approximations algorithm timing	28
Table 3.	Reverse successive approximations algorithm timing	31
Table 4.	Bit current values in the sub-binary DAC	39
Table 5.	Characteristics of the M/3M ladder	63
Table 6.	Characteristics of the M/2.5M ladder	65
Table 7.	2^+ resistor implementation	67
Table 8.	M/2^+M test-chip ladder characteristics	74
Table 9.	M/2^+M current division measurement	74
Table 10.	Calibration table for the example of figure 55	84
Table 11.	Characteristics of the two-stage Miller operational amplifier	96
Table 12.	Closed-loop and open-loop offset measurement	103
Table 13.	Compensation currents for worst-case and Monte Carlo	131
Table 14.	Typical specifications of a current measurement microsystem	172
Table 15.	Combined modulation scheme	176
Table 16.	Demodulation schemes	177
Table 17.	External signal, reference signal and noise levels	184
Table 18.	Sensor and coil characteristics	201
Table 19.	Characteristics of the DDA	206
Table 20.	Characteristics of the operational amplifier	208
Table 21.	External signal demodulation intermediate results	216
Table 22.	Reverse modulation scheme	228

Table 23.	Reverse demodulation schemes	228
Table 24.	Multiplexed modulation scheme	229
Table 25.	Multiplexed demodulation scheme	230
Table 26.	Capacitor values in the reference demodulator	231
Table 27.	Pin functions	233
Table 28.	Demodulator output for $B_{ext} = 0$	235
Table 29.	Demodulator output for negative B_{ext}	236
Table 30.	Demodulator output for positive B_{ext}	237
Table 31.	Microsystem characteristics	240

Chapter 1

Introduction

1 CONTEXT

Ever since the invention of the transistor in the late 50's, its fabrication technology has been evolving, allowing the device integration in a continuously shrinking area. High-performance integrated analog systems have always been difficult to design. Sometimes, calibration is used to gather the extra performance that the analog devices cannot provide intrinsically. But the evolution of the manufacturing technology renders even basic analog systems difficult to design today. With the size reduction, the intrinsic precision of the components degrades. In parallel, the supply voltage decreases, limiting the topologies which can be used. Many modern technologies are specifically suited for pure digital circuits, and some analog devices, like capacitors, are not available. In these conditions, analog design is a challenge even for experienced designers.

To relieve the extreme design constraints in analog circuits, digital calibration becomes a must. It allows a low-precision component to be used in high-performance systems. If the calibration is repeated, it can even cancel the effect of temperature drift and ageing.

The digital calibration is compatible with the evolution of fabrication technologies, which ever more facilitates the integration of digital solutions at the cost of a dramatic reduction of analog performances. Thanks to the reduction of the size of digital devices, even complex digital calibration solutions can be integrated and become a viable alternative to intrinsically precise analog designs.

Digital calibration allows to realize high-performance analog systems with modern technologies. This enables pure analog designs to be implemented even in fully digital processes. In existing mixed-signal designs, the full system realization also becomes possible with technologies providing higher integration density. Finally, because circuit performances rely on digital calibration, retargeting is simplified. The digital blocks can be synthesized automatically, whereas only a limited design effort is invested in the analog circuit.

2 OBJECTIVES

The first objective of this book is to provide a general methodology for the digital calibration of analog circuits. It ranges from the analog circuit analysis (to identify how imperfections are detected) to the implementation of the compensation. It presents systematic means for performing the compensation based on general correction blocks and algorithms. The opportunity of performing regular calibration is also analyzed, and a classification of analog systems allowing or disallowing this feature is developed. Finally, simulation tools permitting the verification of the efficiency of the calibration are presented.

The second objective is to use the defined methodology for correcting the imperfections of existing circuits. In this book, the application of the compensation technique and circuits to three different systems is proposed: a high-precision digital-to-analog converter, a SOI (silicon on insulator) 1T DRAM (single-transistor dynamic random access memory), and a Hall sensor-based microsystem for current measurement.

3 COMPENSATION METHODOLOGY

The compensation methodology is based on current-mode sub-binary radix converters used in conjunction with successive approximations algorithms. A complete analysis of an efficient implementation of sub-binary converters using MOS transistors is performed. In particular, it is demonstrated that these very low-area $M/2^+M$ converters can achieve arbitrarily high resolutions, which is advantageous to perform high-precision calibrations.

An adaptation of the compensation methodology to continuous-time processing systems is also studied. In particular, a way of using an adapted successive approximations algorithm and compensation converter which produce unity up and down compensation steps is presented.

4 APPLICATIONS OF THE COMPENSATION METHODOLOGY

The sub-binary converters are intrinsically non-linear and their direct use as conventional digital-to-analog converters is impossible. However, using two special calibration and radix conversion algorithms, this limitation is removed and the realization of high-precision DACs becomes possible, even with very low-precision components used in sub-binary converters.

The second application is a SOI 1T DRAM, for which an automatic reference calibration technique is proposed. Using the proposed compensation methodology, a sub-binary DAC controlled by a successive approximations algorithm generates the current reference necessary to read the memory. The reference compensates various circuit imperfections together, from the sense amplifier offset to the statistical dispersion of the memory cell currents.

The most important application of the digital compensation methodology is a current measurement microsystem based on a Hall sensor. Until now, the performances of current measurement ASICs have been highly limited by the sensitivity drift of integrated Hall sensors. A novel continuous sensitivity calibration technique is proposed, based on the digital compensation methodology. It combines chopper and autozero techniques, along with all the circuits and algorithms proposed in the first part for the general correction methodology.

5   BOOK ORGANIZATION

Chapter 2 is an introduction to common compensation techniques. The chopper and autozero techniques are presented, and the conditions of their use in continuous and sampled systems is discussed. Finally, both techniques are compared and a classification is performed.

Chapter 3 presents the digital compensation algorithm (successive approximations), and the current-mode sub-binary $M/2^+M$ digital-to-analog converters which are especially well-suited for digital compensation by current injection. Other sub-binary structures are also presented and compared. Finally, the special calibration and radix conversion algorithms, allowing the use of sub-binary converters as conventional DACs, are presented.

Chapter 4 proposes a complete digital compensation methology which allows the correction of circuit imperfections using the circuits and algorithms of chapter 2. The presentation includes specific simulation tools for automatic digital compensation. The application of the methodology to the SOI 1T DRAM reference calibration is presented.

Chapter 5 introduces a new sensitivity calibration technique for Hall microsystems, based on the methodology and circuits of chapters 3 and 4. After an introduction to Hall sensors and the state of the art in Hall sensor-based microsystems, the principle of the calibration technique is explained. The system-level issues are presented and the solutions explained.

Chapter 6 details the implementation of a complete Hall microsystem for current measurement using the sensitivity calibration technique proposed in chapter 5. Each block is presented, and the simulated and measured performances discussed.

Chapter 7 concludes this book by highlighting the most important results and proposing future improvement possibilities.

Chapter 2

Autocalibration and compensation techniques

This chapter presents techniques which are commonly used to compensate or hide imperfections of analog circuits. Some of them, like chopper modulation, use mostly analog circuitry to remove a disturbing effect. Others, like successive approximations, extensively use digital correction algorithms to trim analog components or circuits. First, the mostly used techniques are presented. Then, their performances are examined and a classification is made.

1 INTRODUCTION

The design of analog circuits is rendered difficult by the imperfections imparted by the manufacturing process to the component values. Physical parameters (e.g. oxide thickness, physical dimensions, doping profile) are subject to variations due to instabilities of the fabrication technology, and they reflect on component parameters. The best achievable tolerance of individual component values thus depends on the accuracy of the manufacturing process, and cannot be reduced below a minimum level.

Fortunately, analog design rarely relies on the *absolute value* of single components, but rather on *relative values* of several components. The relative values can be made arbitrarily close, i.e. with small tolerances, by using appropriate design techniques like matching. Thus, high-precision circuits can be realized even with poor manufacturing processes.

2 MATCHING

The most common technique for improving the precision of analog blocks is matching. If the layout of pairs/sets of components is performed carefully following the rules presented below, the statistical dispersion of their values can be reduced.

2.1 Matching rules

The following rules should be applied for optimum matching of integrated components [1]:

1. Same structure
2. Same temperature
3. Same shape, same size
4. Minimum distance
5. Common-centroid geometries
6. Same orientation
7. Same surroundings
8. Non minimum size

When designing pairs/sets of components using these rules, one makes them all as similar as possible. Furthermore, as the components are split and mixed appropriately (common-centroid), they are statistically affected in a similar manner by external (e.g. temperature) and intrinsic (e.g. doping) parameter variations.

2.2 Matching parameters

If the rules presented in section 2.1 are correctly applied, the dispersion of the component values becomes an inverse function of the *area* occupied by the devices [2][3][4]. This means that by increasing the size of the features and by applying rigorously the matching rules, the relative mismatch of the device pairs/sets is reduced. The general model that describes the dependence of the matching of a parameter P on the area of two devices with area $W \cdot L$ is:

$$\sigma^2(P) = \frac{A_P^2}{W \cdot L} \qquad (2.1)$$

where A_P is the process-dependent matching parameter describing the area dependence. This model is applicable to capacitors, resistors, MOS transistors, etc.

The statistical dispersion is inversely proportional to the area of the device. Consequently, in order to achieve a given matching precision, one has to design components larger than the limit that is calculated using equation 2.1. Obviously, the designer faces an important trade-off between precision and circuit area when using only matching properties. But there are also other techniques that allow for increasing the precision of poor circuit elements. Instead of focusing on building high-precision devices, one can build low-precision components and try to *adjust* them or *compensate* for their imperfections later on. There are wide varieties of such techniques, each one having its specific application fields. The new trade-off is then between the matching effort and the use of one or a combination of these compensation techniques. This chapter presents some of them, focusing on the additional circuitry needed to implement them and on the alternative design choices.

3   CHOPPER STABILIZATION

Many imperfections of operational amplifiers, e.g. 1/f noise and offset, are low-frequency or even DC. The idea of chopper stabilization [5][6] is to transpose the signal to a higher frequency where the effect of 1/f noise (and offset) is negligible, to amplify the modulated signal, and finally to demodulate the amplified signal back to the baseband.

3.1   Principle

Figure 1 presents a functional schematic of a chopper amplifier.

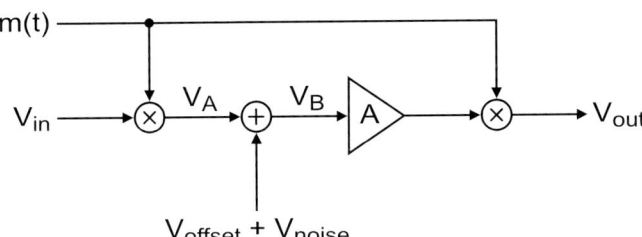

Figure 1.   Functional chopper amplifier

A modulation signal m(t) periodically changes the polarity of the input signal V_{in}. The amplifier block A is ideal, having an infinite bandwidth and neither offset nor noise. However, an equivalent input offset V_{offset} and noise V_{noise} are added to the input V_A of the amplifier, generating an equivalent imperfect input signal V_B for the ideal amplifier. The amplified signal is demodulated by sign changes using the same signal as for input modulation, resulting in the system output V_{out}.

3.2 Analysis

Figure 2 presents an analysis of this chopper amplifier in the time domain, whereas figure 3 displays the frequency analysis.

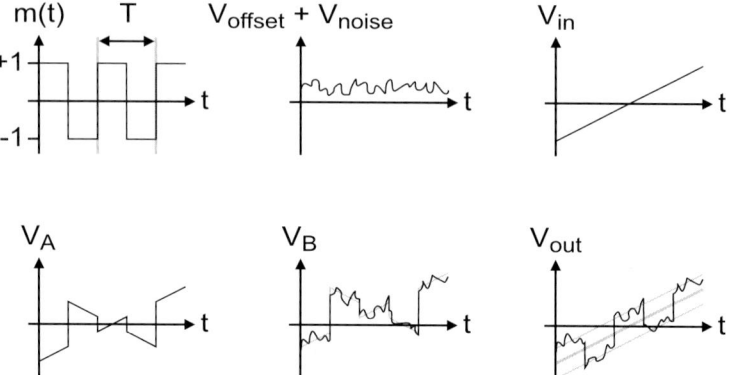

Figure 2. Temporal analysis of a chopper amplifier

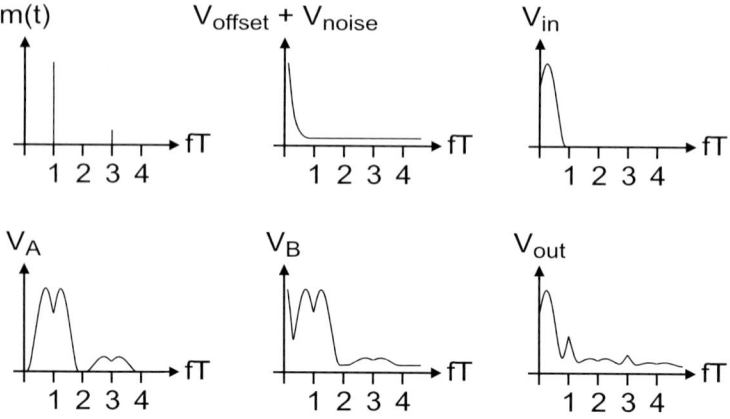

Figure 3. Frequency analysis of a chopper amplifier

In this functional system, the modulation signal m(t) is a square wave with a period T that is applied to both the modulator and demodulator. V_{in} is a band limited signal with frequency components up to at maximum 1/T. If this is not the case, the higher frequencies are aliased in the baseband, which is undesirable.

The modulation changes the sign of the amplifier input periodically, which corresponds in the frequency domain to a shift of the spectrum to the odd harmonics of the modulation signal. This point is the key of the performances of a chopper amplifier. Indeed, the imperfections that are added to the shifted spectrum have important low-frequency components (offset and 1/f noise), whereas they are significantly lower at the frequencies where the signal is shifted. Ideally, the chopper frequency is chosen to be higher than the corner frequency of the 1/f noise in order to add only white noise to the signal.

Once the signal V_B is amplified, it is brought back to the baseband by the demodulator, which effectuates exactly the same operation as the input modulator. The effect is to shift the signal back around DC and even multiples of the chopper frequency, whereas the 1/f noise and offset are located at the odd harmonics. In the time domain, this signifies that the mean value is the amplified signal, whereas the modulated component is the offset.

Obviously, the output signal V_{out} cannot be exploited as is. The signal is correctly present in the baseband, but the higher frequency components should be removed. For this reason, the output of chopper amplifiers is usually low-pass filtered by an additional stage.

3.3 Implementation

To simplify the realization of a chopper amplifier, it is advantageous to use differential inputs and outputs for the amplifier. Indeed, since the inputs and the outputs of the amplifier are differential, changing their polarity is done simply by crossing the positive and negative lines. Such a fully-differential system is presented in figure 4.

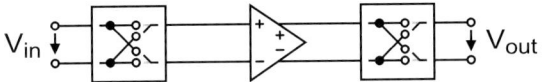

Figure 4. Fully differential chopper amplifier

A schematic of a practical implementation of the modulator and demodulator is the circuit presented in figure 5. Four cross-coupled switches, connected to the modulation signal and its complement, are used for this purpose. When ϕ is active, the input signals are straightly transmitted to the output. When ϕ is inactive, the signals are crossed.

The switches in figure 5 can be realized as CMOS transmission gates, as presented in figure 6. The transmission gate consists of two complementary NMOS and PMOS transistors, which are controlled by complementary signals ϕ and $\bar{\phi}$. The circuit acts as a switch driven by ϕ. It has the advantage over the single-transistor switch of presenting a low-impedance between its terminals A and B, whatever the voltages in both these nodes are.

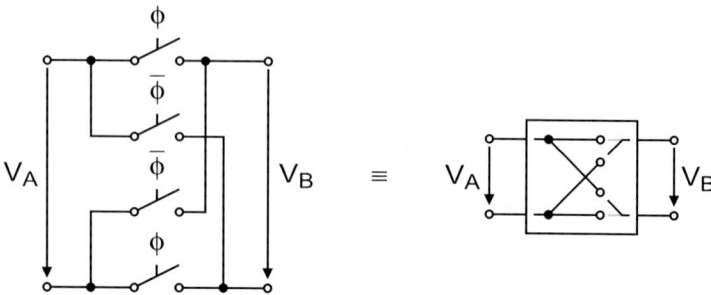

Figure 5. Implementation of a modulator/demodulator using cross-coupled switches

Figure 6. CMOS transmission gate

Using this approach, the implementation of the modulator and demodulator is simple. However, it implies the use of differential inputs and outputs for the amplifier, which is neither practical nor desirable always. Differential inputs are usually available, since most amplifier implementations rely on a differential pair as first stage.

If the amplifier has only one single output, the sign change in the demodulator is more difficult to realize. Figure 7 presents an example of circuit implementing the required function. When ϕ is active, the input signal is directly fed to the output. When ϕ is inactive, the amplifier changes the sign of the input since its gain is designed to be -1.

The main drawback of this solution is the difficulty to obtain precisely the -1 gain, because it depends on the quality of the matching of the two resistors. A second problem arises from the delay introduced by the additional amplifier, making the circuit asymmetrical for both phases. Finally, the imperfections of the additional amplifier, such as offset and noise, degrade the overall system performance. In this example, this is not problematic if the chopper amplifier gain A is high, because the input-referred offset and noise of the amplifier in the demodulator are divided by A.

As one can see, single input and/or output chopper amplifiers are less straightforward to design. In some specific applications however, these circuits are more suitable than differential topologies.

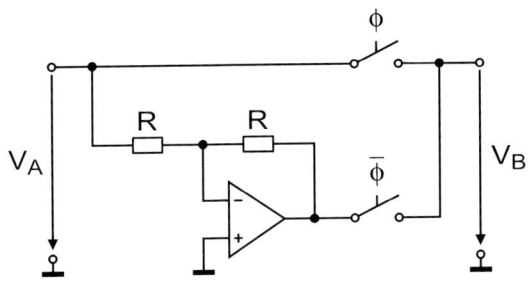

Figure 7. Demodulator for single output chopper amplifier

4 AUTOZERO

Autozero is another common technique used to minimize offset and 1/f noise in amplifiers. The main idea [7] is to first sample the undesired effect and then to subtract it during the second phase when the input signal is processed by the imperfect amplifier.

4.1 Principle

Figure 8 presents the principle of an autozero amplifier [6], which is also applicable to comparators. The amplifier A is ideal, the real amplifier noise and offset being represented by the voltage source connected to the positive input.

During the first phase, the amplifier is disconnected from the input signal by switch S_{in} and the offset V_O and noise V_N voltages are sampled[1] on capacitor C_{AZ} across switch S_{FB}:

$$V_C = \frac{A}{1+A} \cdot (V_O + V_N) \cong (V_O + V_N) \qquad (2.2)$$

---

[1]. The assumption is made that the open-loop gain A of the amplifier is much larger than 1, which is correct in most cases.

During the second phase, the amplifier is used in its normal processing configuration. The autozero capacitor is connected so that it cancels the effect of the parasitic voltage source.

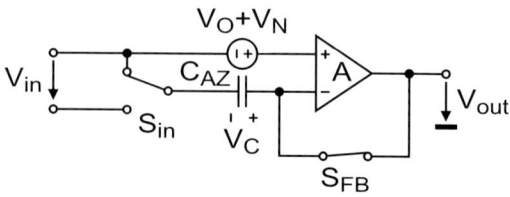

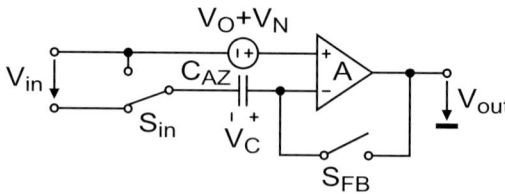

Figure 8. Autozero amplifier principle
Top: Sampling phase; Bottom: Processing phase

4.2 Analysis

The circuit presented in figure 8 does not correspond to the usual implementation of an autozero amplifier. There are actually more convenient ways to realize the same functionality.

For this example topology, the offset and the noise voltages are stored in a capacitor. Usually, their magnitude is smaller than 10 mV. The switches are implemented as transmission gates (see figure 6), or simply as single transistors. Therefrom arises one important problem due to the imperfections of MOS switches. When such a switch is opened, the well-known charge injection [8] phenomenon occurs. The charges that remain in the channel of the transistor flow into its source and drain, causing an undesired voltage variation across the sampling capacitor.

There are techniques for reducing charge injection. One can take special care when designing the switches, reduce the slope of the clock signal that triggers the switch, increase the size of the capacitor, etc. Nevertheless, this effect cannot be cancelled completely and it is difficult to manage when the signal level to which it is compared is small, like the 10 mV stored in C_{AZ}. It is thus advantageous to increase the voltage V_C in order to limit the effect of charge injection. But this cannot be achieved using this topology.

Chapter 2: Autocalibration and compensation techniques

Figure 9 presents another autozeroed amplifier, for which charge injection is less problematic. In this circuit, the amplifier has an additional input Z used for offset nulling. During the offset sampling phase, the feedback loop including an auxiliary gain/attenuation stage A' and the sample and hold circuit adjusts the nulling signal so that V_{out}, and consequently also the amplifier offset, become null. The compensation value is then held during the second phase when the circuit is in the normal amplification mode.

The advantage of this technique over the one described before (figure 8) is that the signal level in the offset storage capacitor can be arbitrarily increased. It depends on the gain of the nulling input of the amplifier.

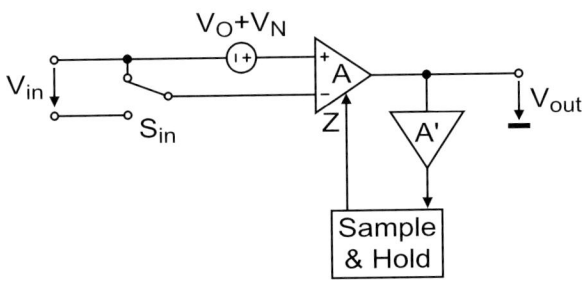

Figure 9. Analogically compensated autozero amplifier

Using the same topology, one can perform the offset cancellation in the digital domain with the circuit of figure 10.

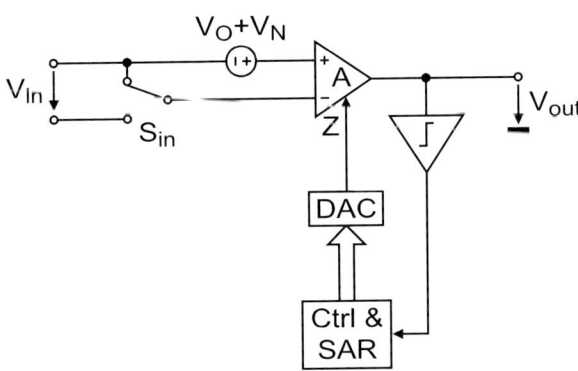

Figure 10. Digitally compensated autozero amplifier

The sample and hold circuit is in this case a digital-to-analog converter, which is controlled by a digital circuit performing successive approximations (see also section 3 in chapter 3). Since the information necessary to decide whether the value of the DAC is too large or not is the sign of V_{out}, the gain/attenuation stage of the previous fully analog circuit is here a comparator.

The advantage of the digital approach is that it allows the compensation information to be stored during an unlimited time, whereas the analog sample and hold memory used in the analog solution needs periodic refreshing. However, this is an advantage only if one considers one-time calibration. In the case of periodic offset nulling, which is interesting because it also cancels 1/f noise, the attractiveness of the digital solution diminishes. On the other hand, the problem of charge injection, which is one of the limitations of the analog solution, is not present if a digital compensation is implemented. In this case, the quality of the compensation depends on the resolution of the DAC.

4.3 Noise

If the autozero process is carried out at a sufficiently high frequency, it also reduces the low-frequency 1/f noise. A detailed analysis of the effect of autozero on noise can be found in [6]. This section presents only the most important results.

The resulting noise after autozero S_{AZ} has two components: the remaining baseband noise below the autozero frequency f_{AZ} (period T_{AZ}), and a foldover component which is aliased to the baseband because of the noise sampling that is done by the autozero process:

$$S_{AZ}(f) = S_{base}(f) + S_{fold}(f) = |H_0(f)|^2 S_N(f) + \sum_{\substack{n = -\infty \\ n \neq 0}}^{n = \infty} |H_n(f)|^2 S_N(f - nf_{AZ}) \quad (2.3)$$

with

$$|H_n(f)|^2 = \begin{cases} \left(1 - \dfrac{\sin(2\pi fT_{AZ})}{2\pi fT_{AZ}}\right)^2 + \left(\dfrac{1 - \cos(2\pi fT_{AZ})}{2\pi fT_{AZ}}\right)^2 & n = 0 \\ \dfrac{\sin(\pi fT_{AZ})}{\pi fT_{AZ}} & n \neq 0 \end{cases} \quad (2.4)$$

In equations 2.3 and 2.4, S_N and S_{AZ} are the power spectral density of the noise without and with autozero respectively. S_{base} and S_{fold} are the power spectral densities of the baseband and the foldover noise contributions. H_0 is the noise transfer function in the baseband, and H_n ($n \neq 0$) is the transfer function of the foldover of the nth harmonic due to the autozero modulation. These functions are plotted in figure 11, with a normalized x-axis.

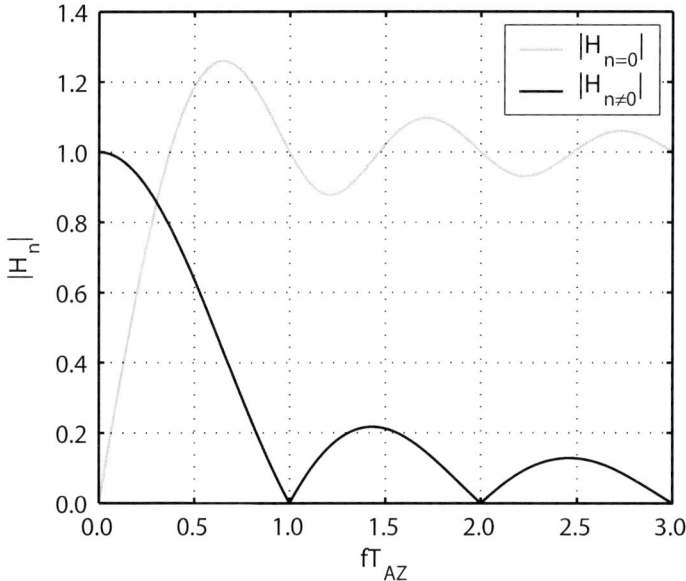

Figure 11. Autozero baseband and foldover noise transfer functions

On the one hand, the baseband noise is attenuated by a high-pass like function, which means that the DC offset and low-frequency noise is strongly attenuated by the autozero technique. On the other hand, the foldover components increase the white noise floor.

The amplitude of this foldover component depends on the bandwidth of the amplifier. Let's consider the common case where the amplifier has white and 1/f noise with a corner frequency at f_k and a white noise floor power spectral density S_0. Let's further assume that the amplifier performs a first-order low-pass filtering of this noise with a dominant pole at f_c. The noise power spectral density is in this case:

$$S_N(f) = S_0\left(1 + \frac{f_k}{f}\right)\left(\frac{1}{1 + \left(\frac{f}{f_c}\right)^2}\right) \qquad (2.5)$$

The total foldover noise is [6]:

$$S_{fold}(f) = S_0 \left\{ (\pi f_c T_{AZ} - 1) + 2 f_k T_{AZ} \left[1 + \ln\left(\frac{2}{3} f_c T_{AZ}\right) \right] \right\} \frac{\sin(\pi f T_{AZ})}{\pi f T_{AZ}} \quad (2.6)$$

Replacing equations 2.4, 2.5 and 2.6 in 2.3 allows to calculate the resulting noise level S_{AZ} after autozero. Figures 12 and 13 present the normalized plots of the resulting S_{AZ} and original S_N noise power spectrum for two different amplifier bandwidths: 5 and 50 times the autozero frequency respectively. On both plots, the white noise floor is located at 0 dB on the y-axis.

The resulting noise spectrum after autozero in the baseband (below 10^0 on the x-axis) is almost flat and clearly shows that the 1/f noise is cancelled. On the other hand, the amplitude of the noise floor in the baseband is higher than the white noise floor. This is caused by the foldover of both white and 1/f noise. The total amount of foldover noise depends on the amplifier bandwidth: The smaller the bandwidth, the less noise is aliased in the baseband. For this reason, f_{AZ} should not be chosen much smaller than f_c. On the other hand, to allow sufficient settling time for the voltage on the sampling capacitor, one should choose [9]:

$$f_{AZ} \leq \frac{f_c}{5} \quad (2.7)$$

Figure 12 shows the result of the choice of an optimal f_c/f_{AZ} ratio, whereas figure 13 corresponds to a suboptimal one.

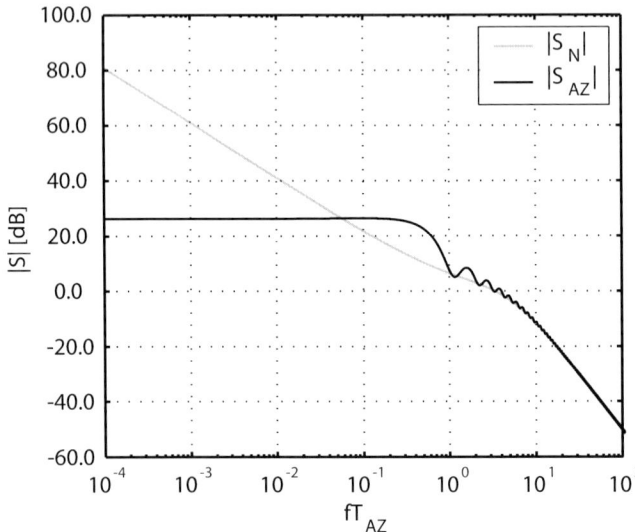

Figure 12. Resulting noise with autozero and small amplifier bandwidth ($f_c = 5 f_{AZ}$)

Chapter 2: Autocalibration and compensation techniques

Concerning the effect of the 1/f corner frequency f_k on the resulting noise, let's examine equation 2.6 again. If $f_c/f_{AZ} = 5$, the middle term of the equation (between the braces) becomes:

$$m(f_k) = 5\pi - 1 + 2\left(1 + \ln\frac{10}{3}\right)\frac{f_k}{f_{AZ}} \cong 14.71 + 4.41\frac{f_k}{f_{AZ}} \qquad (2.8)$$

This function has a multiplicative effect on the overall resulting noise function S_{fold}. Figure 14 shows a normalized plot of $m(f_k)$.

In order to keep the noise level low, it is advantageous to choose:

$$f_{AZ} \geq f_k \qquad (2.9)$$

However, it is not necessary to have $f_{AZ} \gg f_k$, since this also means unnecessarily increasing the bandwidth of the amplifier (equation 2.7).

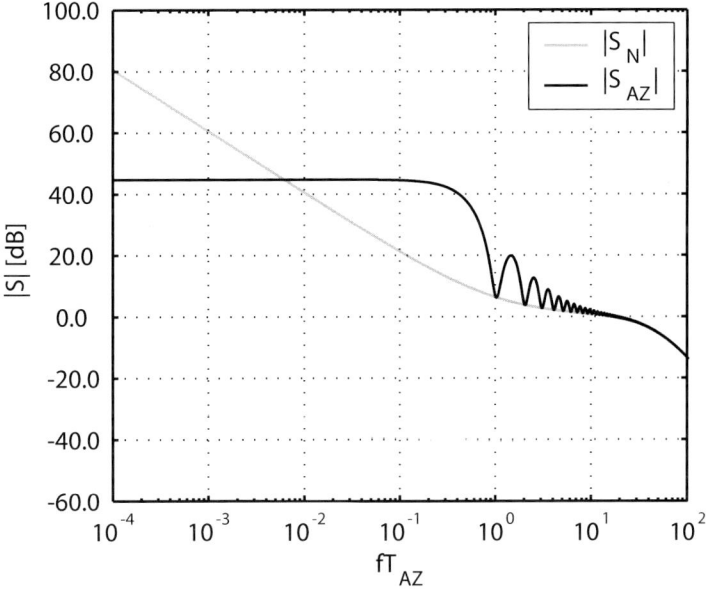

Figure 13. Resulting noise with autozero and large amplifier bandwidth
($f_c = 50 f_{AZ}$)

Finally, combining equations 2.7 and 2.9 gives the optimal design guideline for autozero amplifiers:

$$f_k \leq f_{AZ} \leq \frac{f_c}{5} \qquad (2.10)$$

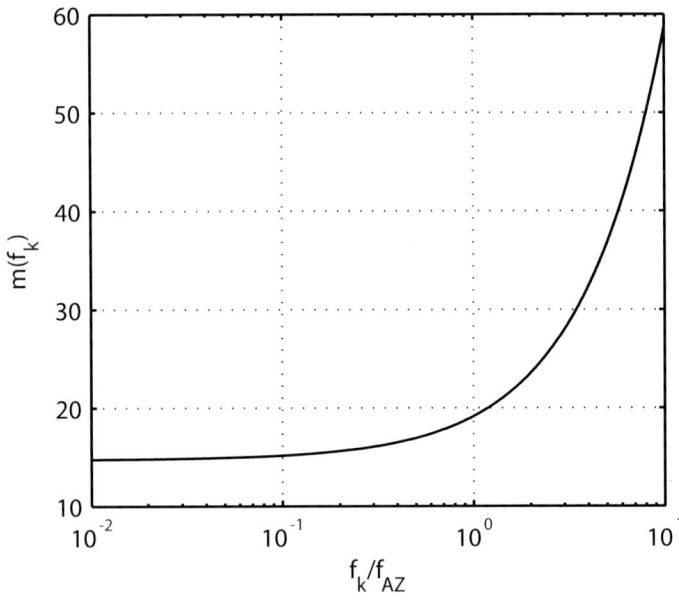

Figure 14. Effect of the 1/f corner frequency on the resulting noise

5 CORRELATED DOUBLE SAMPLING

Correlated double sampling (CDS) is similar to the autozero technique presented in section 4. The main difference is the following: With autozero, the noise is first sampled, and then the amplifier performs continuous-time output, subtracting the sampled noise value. In the case of CDS on the other hand, the second phase samples the signal. The subtraction of the two sampled values allows to remove the offset and noise from the signal, exactly as with autozero. The first application of CDS was in CCD sensors [10], but the same technique can be used in any sampled signal processing system.

CDS is also similar to autozero in terms of noise and offset cancellation: There is a cancelling of 1/f noise and offset, and a foldover component which increases the noise floor in the baseband.

6 PING-PONG

As explained in section 5, the autozero and CDS techniques can be used in sampled data systems. However, when a continuous-time output is required, they are not applicable directly because the system must be disconnected from the signal path during offset and noise sampling. This problem can be overcome by using the ping-pong technique.

The principle is to duplicate the signal processing circuit, in this case an amplifier. One of them is calibrated, while the other is amplifying the input signal [11]. When calibration is done, the role of both amplifiers is reversed: The just calibrated amplifier is used for signal amplification, and the second is calibrated. The same procedure goes on and on, hence the name "ping-pong".

Figure 15 presents an improved ping-pong technique [12] that reduces the glitches during the transition from one amplifier to another. On the left, amplifier A amplifies the signal and A' is under calibration. During the transition phase in the middle, amplifier A' copies the amplified output of A. This prevents the system from generating an important transient spike as if A' directly replaces A. On the right, the roles of both amplifiers are exchanged: A' processes the signal and A is calibrated.

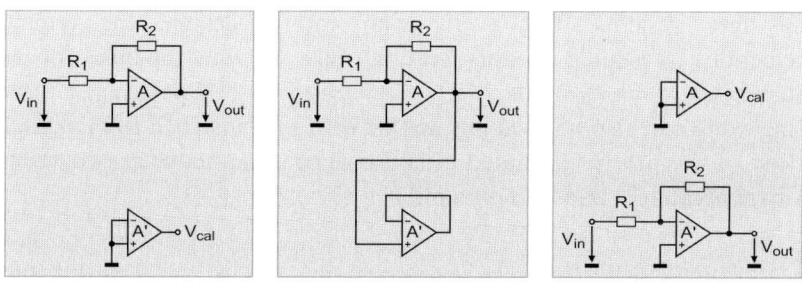

Figure 15. Ping-pong amplifier system
Left: A' calibration; *Middle*: Transition; *Right*: A calibration

Even if the transition between both amplifiers is done cautiously, there still remain two undesirable effects as presented in figure 16 [13]. First, the amplitude of the spike at the instant of swapping t_{swap} is reduced by using an intermediate phase, but not removed completely. Second, since the offset voltages $V_{offset, A}$ and $V_{offset, A'}$ of both amplifiers A and A' are not exactly equal when the amplifiers are swapped, the output signal is slightly shifted after the transition by:

$$\Delta = V_{offset, A'}(t_{swap}) - V_{offset, A}(t_{swap}) \qquad (2.11)$$

The amplitude of the step Δ can be reduced by increasing the precision of the offset correction. But if the time interval between two successive swaps is long, the offset of the active amplifier can drift, or the low-frequency noise can have an equivalent influence. In this case, Δ is increased even if the offset of the other amplifier is calibrated precisely.

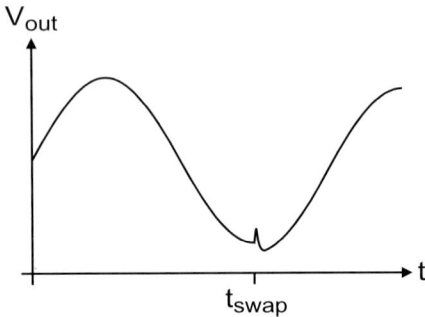

Figure 16. Operational amplifier swapping

Ping-pong allows to use components that need offline calibration in continuous-time systems. But even using careful design, the remaining imperfections presented in this section may be unacceptable for some applications.

Ping-pong can also be used in sampled data systems [14]. In this case, the insertion of the newly calibrated circuit can be done between two samples, without generating a transient disruption.

7 OTHER TECHNIQUES

The preceding sections present a limited overview of compensation and calibration techniques, as the emphasis is voluntarily put on the most widespread solutions for calibrating analog circuits, like operational amplifiers. Obviously, there are many other systems, like digital-to-analog and analog-to-digital converters, where specific techniques are used for improving performance of analog circuits.

Some of these techniques have the same background principle as others presented in this chapter. Dynamic element matching [15] for instance, implements the matching technique presented in section 2 in the time domain instead of in the area domain: The errors of the components are averaged by using different devices at different instants.

On the other hand, there are numerous other techniques used in data converters which cannot be applied to other analog circuits because they are specific. Examples of techniques used in ADCs and DACs can be found in [16].

Chapter 2: Autocalibration and compensation techniques 21

8 CLASSIFICATION

The techniques presented in this chapter all have their specific application field. In many cases, several techniques can be considered to solve the same problem. In some situations, they can even be combined to benefit from the advantages of each of them, without necessarily suffering from the drawbacks. For instance, it is possible to combine chopper, autozero and ping-pong to realize an amplifier [17] which has at the same time low offset, low noise in the baseband (thanks to chopper) and high bandwidth (because of autozero).

To facilitate the choice between different solutions, this section offers a comparison of their advantages, drawbacks and application possibilities. Table 1 presents a condensed comparison of chopper and different variants of autozero.

Table 1. Characteristics of the compensation techniques

Compensation technique	Signal		noise & BW		Calibration	
	Sampled	Continuous	Low baseband noise	High bandwidth	Low-frequency	High-frequency
Chopper	✓	✓	✓			✓
Analog autozero	✓			✓		✓
Digital autozero	✓			✓	✓	✓
Autozero & Ping-Pong	✓	✓		✓		✓

The first two columns reflect that both chopper and autozero are directly applicable to sampled signal processing systems like switched-capacitor circuits, analog-to-digital and digital-to-analog converters, etc. Chopper can also be used in continuous-time applications, while autozero is only usable there in conjunction with ping-pong, because the calibration process requires removing the circuit from the signal path.

If very low baseband noise is required, chopper is preferable because contrary to autozero, there is no noise folded into the baseband. If the noise level is not critical, autozero is also suitable (see section 4.3 for details). In terms of 1/f noise removal, they are both comparable.

On the other hand, when a high signal bandwidth is necessary, autozero is much more adapted because it keeps the signal in the baseband instead of modulating it at higher frequencies.

The two last columns of table 1 present the characteristics regarding the compensation frequency. If the system is calibrated frequently, all solutions are adapted. If on the other hand the calibration is performed at long time intervals, the autozero using digital compensation circuitry is much easier to implement. This is due to the fact that analog memories have a limited retention time and need to be refreshed periodically.

There exist analog storage techniques like floating gate which allow extremely long retention times [18], but they imply extra processing steps, specific devices and/or high programming voltages [19]. On the other hand, storing digital information over a long period of time is straightforward. The only limitation is the data loss due to power off.

Another approach is factory trimming, which is one time programming. It can compensate for initial imperfections, but does not correct drifts due to circuit ageing or temperature variations. This approach is somewhat different from chopper and autozero, but it can in some cases be sufficient.

The compensation frequency depends on the factors that affect the variations of system parameters. If the circuit is very stable, a one time factory trimming can suffice. If the characteristics change with the age of the circuit, a single calibration at power-up can be considered. If they change with temperature, calibration can be performed periodically at low frequency (in the Hz range). Finally, if the variation is faster, like 1/f noise for instance, the compensation must also be performed at higher rates.

9 CONCLUSION

There are various techniques to improve the performances of analog circuits. Although they all have specific applications, the digital solutions generally seem better adapted to the long-term evolution of manufacturing processes. With the ever more shrinking of devices, it becomes more and more difficult to design intrinsically high-performance analog circuits. If digital correction circuits are available to compensate for the imperfections of analog components, it will still be possible to design high-performance circuits with future deep sub-micron fabrication processes. Furthermore, if the digital correction circuits and algorithms are systematic, it will be possible to automate their design and allow the designers to focus their attention on the analog blocks.

Chapter 3

Digital compensation circuits and sub-binary digital-to-analog converters

This and the next chapter present a complete methodology for digital compensation of analog circuits. In chapter 3, all the necessary circuits are presented. Chapter 4 focuses on the design methodology for digitally compensated circuits. The techniques and circuits presented are based on successive approximations algorithms. After an introduction to the algorithm and its working condition, sub-binary digital-to-analog converters are presented. As successive approximations algorithms neither require monotonic nor linear converters, it is possible to design very high-resolution sub-binary DACs with limited design effort. Different sub-binary topologies are described in this chapter, in particular the current-mode $M/2^+M$ converter.

1 INTRODUCTION

Among the compensation techniques presented in chapter 2, the digital correction is the most versatile. Without requiring major adaptations, it can be used in a large variety of situations to compensate for analog circuit imperfections.

This chapter presents successively all the elements needed for digital compensation. First, the successive approximations algorithm is presented. Then, different structures of digital-to-analog converters that can be used in conjunction with this algorithm are studied and their performance compared.

2 DIGITAL COMPENSATION

To foster understanding, this chapter refers to digital compensation on the basis of the example presented in figure 17, unless otherwise specified. The circuit is the same as in figure 10.

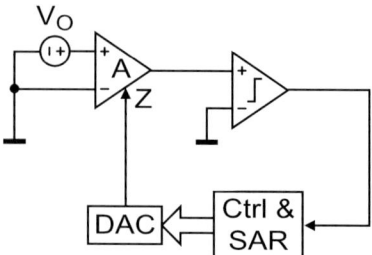

Figure 17. Digital compensation of the offset of an operational amplifier

The sign of the offset voltage V_O of the amplifier A is detected using a comparator connected to its output. The amplifier has an extra analog offset nulling input Z, which is considered positive[1] by convention. It can be a voltage, a current, or another physical quantity, but remains here without unit to keep the presentation as general as possible.

The feedback loop between the output of the comparator and the correction input is the digital compensation circuit. The digital information available at the output of the comparator, namely the sign of the offset, is sufficient[2] to allow complete offset cancellation. If the offset is positive, the correction circuit decreases it by diminishing the Z input value. Conversely, if the sign is negative, Z is increased.

On this basis, it is possible to design a compensation circuit consisting of an adapted algorithm realizing the function described above and a digital-to-analog converter driving the compensation input. This chapter analyses these blocks and presents the simplifications and optimizations that can be made.

3 SUCCESSIVE APPROXIMATIONS

The successive approximations algorithm allows to find quickly a digital value by dichotomic search. By performing a sequence of comparisons, it converges to the desired value. The algorithm is presented here as a means for calibrating circuits. However, it has other well-known applications, in analog-to-digital converters for example [7].

Let's consider that the DAC of figure 17 is an ideal binary-radix digital-to-analog converter with n bits of resolution. It has a digital input bus D, consisting in n digital inputs (d_1, ..., d_n), where d_1 corresponds to the least significant bit (LSB) and d_n to the most significant bit (MSB). The analog output value A

[1.] i.e. increasing the input value Z increases the offset voltage V_O.

[2.] because the relation between Z and V_O is a strictly monotonic function.

Chapter 3: Digital compensation circuits

of the DAC for each bit is $b_1, ..., b_n$. They are binary weighted and perfectly linear, and $b_1 = 1$ arbitrarily. Consequently:

$$b_i = 2^{i-1} \qquad \forall i \in [1...n] \qquad (3.1)$$

The output of the DAC for a digital input word D is:

$$A = \sum_{i=1}^{n} d_i b_i \qquad (3.2)$$

with $d_i = 0$ if bit i is cleared, and $d_i = 1$ if bit i is set.

The input/output characteristics of such an ideal DAC with 4 bits (n = 4) is shown in figure 18. The grey line is the identity function y = x.

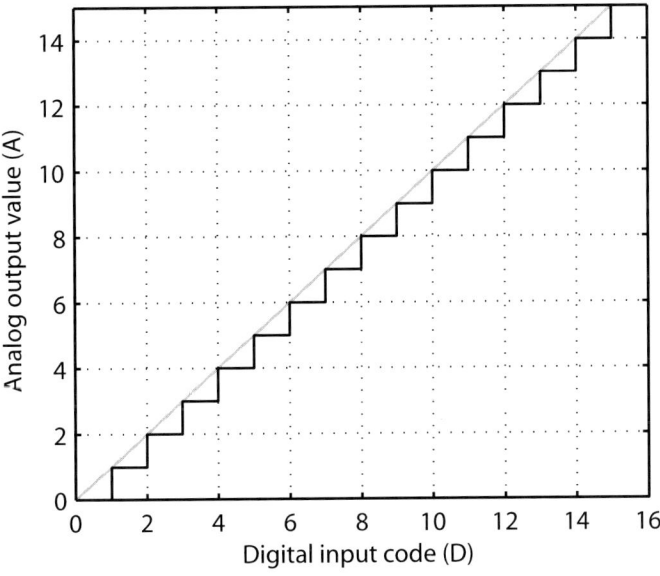

Figure 18. Ideal 4-bits DAC input/output characteristics

3.1 Principle

In the example of section 2, the goal of the algorithm is to find the most suitable digital control value for the DAC which produces the smallest residual offset.

To keep the reflection as general as possible and to simplify the presentation, the offset voltage is not represented directly. Instead, an equivalent compensation value Z_0 is used, which corresponds to the necessary correction value generated by the DAC in order to completely cancel the offset. This is strictly equivalent since there is a direct relation between the nulling input Z and the input-referred offset of the amplifier.

In figure 19, Z_0 is set at 10.5 and plotted in grey. This number has voluntarily no unit (see section 2). If the DAC generates a compensation value above Z_0, the resulting offset is positive and the output of the comparator is also positive. If the compensation value is below 10.5, the output of the comparator becomes negative.

To find the DAC code that lies closest to Z_0, the successive approximations algorithm (figure 20) is used.

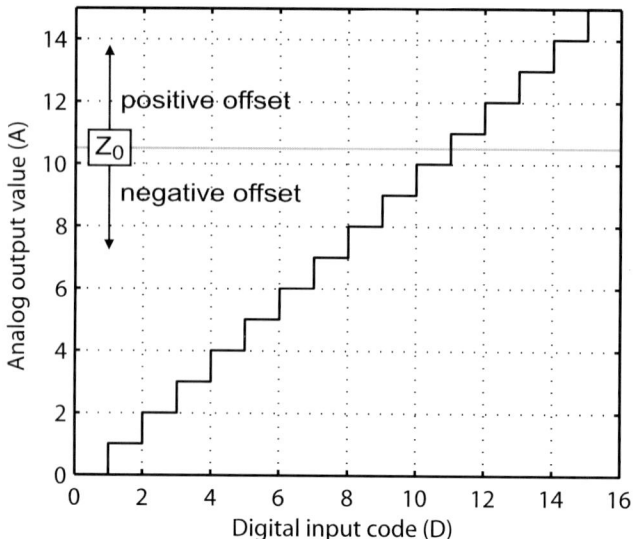

Figure 19. Equivalent offset

```
reset all d_i = 0
for i = n downto 1
    set d_i = 1
    if C_out > 0
        reset d_i = 0
    end if
end for
```

Figure 20. Successive approximations algorithm

Chapter 3: Digital compensation circuits

First, all bits are cleared (D = 0). Then, the bits are tested successively in a loop, starting with the MSB and going down to the LSB. For each bit, the output value of the comparator C_{out} is examined when the bit is set. If the value is negative, this signifies that the remaining offset is negative and that the DAC value has still to be increased. For this reason, the bit is kept. If on the contrary C_{out} is positive, the currently tested code is too high (above Z_0) and the tested bit is reset. The algorithm then performs the same test with the next less significant bit, until reaching the LSB.

Figure 21 shows the analog output value of the DAC versus time (algorithm step). Table 2 shows which comparisons are made and which bits are kept.

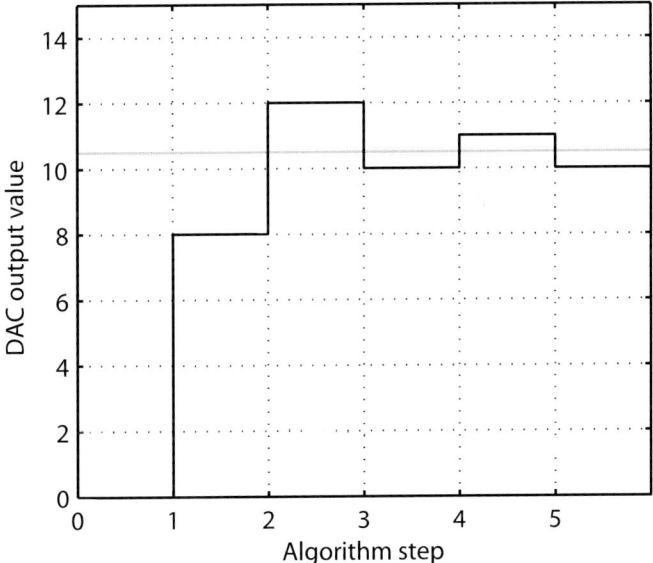

Figure 21. Successive approximations algorithm timing

Table 2. Successive approximations algorithm timing

Step	DAC output value	Comparison result	Comment
0	0	-	reset
1	8	$8 < Z_0$	bit 4 is kept
2	12	$12 > Z_0$	bit 3 is rejected
3	10	$10 < Z_0$	bit 2 is kept
4	11	$11 > Z_0$	bit 1 is rejected
5	10	-	final value

3.2   Working condition

At the end of the successive approximations algorithm, it can be proven that the adjusted DAC output value A_{final} is not more distant than one LSB from the ideal value Z_0:

$$|A_{final} - Z_0| \leq b_1 \qquad (3.3)$$

Furthermore, the final value never exceeds the ideal value:

$$A_{final} \leq Z_0 \qquad (3.4)$$

These properties are the consequence of the construction of the algorithm. Equation 3.4 is valid because the bits elevating the DAC output above Z_0 are systematically rejected. By adapting the algorithm, it is possible to have the converse relation.

Equation 3.3 is true only if the converter has no missing code:

$$b_i \leq b_1 + \sum_{j=1}^{i-1} b_j \qquad \forall i \in [2...n] \qquad (3.5)$$

Chapter 3: Digital compensation circuits

In addition, the value to adjust must be comprised in the DAC full scale:

$$0 \leq Z_0 \leq \sum_{i=1}^{n} b_i \qquad (3.6)$$

Equation 3.5 guarantees that the sum of the less significant bits of any bit b_i in the DAC is not inferior by more than one LSB to the bit itself (b_i). This important condition ensures that if a bit b_i is rejected during a given step of the successive approximations algorithm, the remaining steps using only less significant bits cover the complete range of values up to the just rejected value. If this condition is not satisfied, equation 3.3 is no longer valid.

Let's look again at equation 3.5 and compare it to the values of the weights of the bits in a binary DAC like the one that is used in the preceding examples. In this DAC, equation 3.1 can be rewritten as:

$$b_i = 2^{i-1} = b_1 + \sum_{j=1}^{i-1} b_j \qquad (3.7)$$

This equation is the limit case of inequation 3.5. It means that at each step of the algorithm, the remaining codes are just sufficient to produce the necessary number of values. In other words, the binary digital-to-analog converter has no missing code.

3.3 Reverse successive approximations algorithm

As stated by equation 3.4, the adjusted value of the algorithm is never higher than the ideal value. In some cases, one is interested in the reverse property, i.e. that the adjusted value is never lower than Z_0:

$$A_{final} \geq Z_0 \qquad (3.8)$$

This is achieved by inverting all the bits and the decision in the algorithm. As a result, the reverse successive approximations algorithm (figure 22) is obtained.

```
set all d_i = 1
for i = n downto 1
    reset d_i = 1
    if C_out < 0
        set d_i = 1
    end if
end for
```

Figure 22.   Reverse successive approximations algorithm

Equation 3.3 still applies to this modified algorithm, guaranteing that the maximum distance to the ideal value is 1 LSB.

Figure 23 and table 3 show the execution of this modified algorithm.

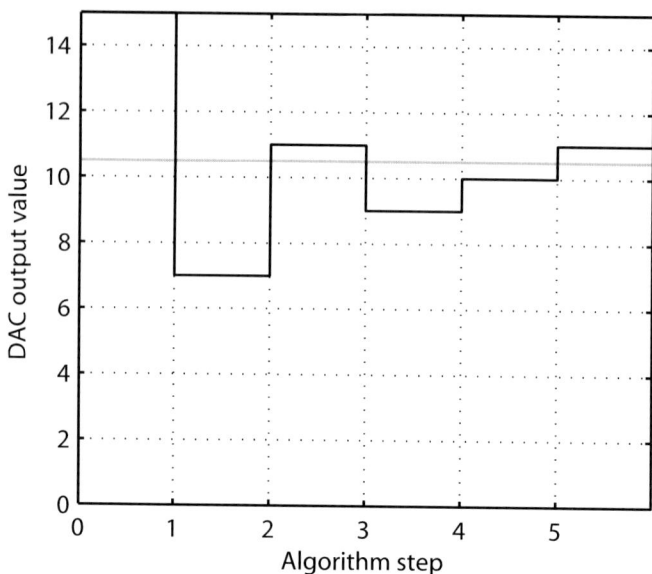

Figure 23.   Reverse successive approximations algorithm timing

Chapter 3: Digital compensation circuits

Table 3. Reverse successive approximations algorithm timing

Step	DAC output value	Comparison result	Comment
0	15	-	reset
1	7	$7 < Z_0$	bit 4 is rejected
2	11	$11 > Z_0$	bit 3 is kept
3	9	$9 < Z_0$	bit 2 is rejected
4	10	$10 < Z_0$	bit 1 is rejected
5	11	-	final value

3.4 Complexity

For a DAC with n bits, the successive approximations algorithm uses n calibration steps (the loop of the algorithm). Compared to the number N of different possible output values, this represents:

$$n = \log_2(N) \tag{3.9}$$

This well-known result shows that the dichotomic search performed by successive approximations is much faster than a systematic testing of all codes, which can be done by replacing the control block of figure 17 by a counter.

4 SUB-BINARY RADIX DACS

4.1 Use of sub-binary DACs for successive approximations

As already discussed in section 3.2, using a binary-radix DAC with a successive approximations algorithm exploits the limit of the working condition. Furthermore, such DACs are difficult to design unless the number of bits is small, because the weight of each bit needs to be precisely set to avoid missing codes and redundancies. For this reason, they usually occupy an important circuit area.

The working condition (equation 3.5) of the successive approximations algorithm does not impose precise bit weights. Whereas missing codes are not allowed, redundancies are not problematic. This can be turned into an advantage, because it allows the use of imprecise converters. By voluntarily introducing redundancies, the risk of missing codes is reduced and the sub-binary converters can be designed using less effort and area without degrading the performance of the algorithm.

Sub-binary converters should systematically be used with successive approximations algorithms. There is no reason for preferring a conventional radix-2 converter.

4.2   Characteristics

In a sub-binary radix DAC, the weights of the different bits are based on a radix R that is smaller than 2:

$$b_i = R^{i-1} \qquad \forall i \in [1...n] \qquad (3.10)$$

Figures 24 and 25 show the input/output characteristics of sub-binary radix DACs with R equal to 1.75 and 1.5 respectively. In both figures, the grey line is the identity function y = x. See also figure 18 for the R = 2 case.

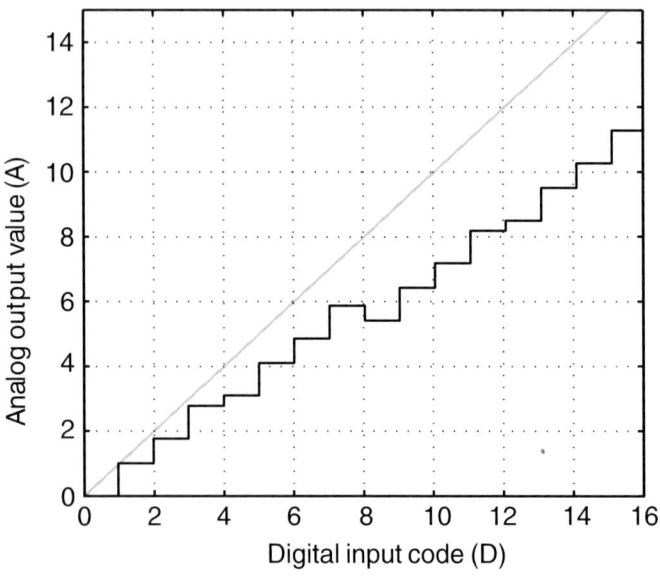

Figure 24.   Input/output characteristics of a radix 1.75 DAC

Chapter 3: Digital compensation circuits

When the radix is decreased, the characteristics move down from the ideal y = x line. The converter becomes non-linear, non-monotonic (some steps are down instead of up) and redundant (the same output code is generated by two or more different input words).

The advantage of the sub-binary radix converter is that it is less likely to have missing codes than a radix-2 converter. Actually, a mismatch problem in a sub-binary converter has the same effect as changing its radix. Provided the radix is sufficiently far from the limit value 2, this does not create missing codes.

The radix has to be chosen carefully to account for variations and to make sure the value 2 is not exceeded. The more variations are expected, the lower the radix must be. The exact choice of the radix depends on the structure of the DAC and the tolerances on its components. This is discussed in sections 5, 6 and 10, which present 3 different topologies of digital-to-analog converters and how they can be made sub-binary.

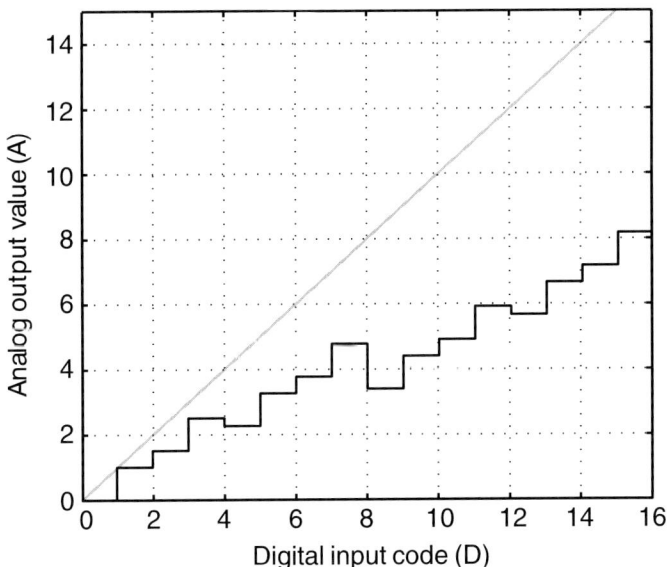

Figure 25. Input/output characteristics of a radix 1.5 DAC

4.3 Resolution

The resolution of a digital-to-analog converter is the ratio between the full scale FS (the range of possible output values) and the LSB value (the smallest difference between two codes). It and can be calculated easily if the radix R is constant:

$$\text{Res} = \frac{FS}{LSB} = \frac{b_1 + \sum_{i=1}^{n} b_i}{b_1} = \frac{R^n}{R-1} \qquad (3.11)$$

The resolution is usually expressed in bits, by calculating the $\log_2$ of the previous equation.

The advantage of code redundancy of sub-binary radix converters is also their slight disadvantage. By creating duplicate codes, the full scale is reduced, and consequently also the resolution. In the example of the radix-1.5 converter (figure 25) for instance, the full-scale of this 4 bits DAC is about 8, which is only half of an equivalent radix-2 converter (figure 18). The 4 bits sub-binary converter has only 3 bits of resolution. The extra bit creates code redundancies.

4.4 Tolerance to radix variations

This section presents a simple design rule for sub-binary converters (equation 3.13). Although it is more restrictive than required, it is easy to translate into a circuit condition.

Since the radix in sub-binary converters is subject to variations due to the tolerances of the components, one can rewrite equation 3.10 as follows:

$$b_i = r_i b_{i-1} \qquad \forall i \in [2...n] \qquad (3.12)$$

where r_i is the "local" radix of stage i.

In this case, a *sufficient but not imperative* condition for the DAC to respect the working condition of the successive approximations algorithm (equation 3.5, is the following:

$$r_i \leq 2 \qquad \forall i \in [2...n] \qquad (3.13)$$

Chapter 3: Digital compensation circuits 35

Eventually, the radix r_i at a given stage can also be slightly higher than 2, but only if the remaining radixes of the less significant bits are sufficiently low (and some of them strictly lower than 2). Taking this parameter into account increases the complexity of the condition, transforming it back into equation 3.5.

Usually, equation 3.13 can easily be translated into a circuit constraint ensuring proper functioning of the converter with the successive approximations algorithm. Furthermore, this constraint is simple to guarantee for all the stages of the converter, even the higher order ones (most significant bits). Adding an extra bit is not more complex at a high or low stage (e.g. adding a 16th bit is not more difficult than adding a 4th). This means that *a sub-binary DAC can be designed with low effort for any given resolution, without increasing circuit complexity.*

Finally, it is not necessary to choose the same radix for all the stages of the converter (i.e. choosing all r_i equal), as long as the local radixes respect the general rule of equation 3.5, or the more restrictive rule of equation 3.13.

5 COMPONENT ARRAYS

The most intuitive manner to compensate for imperfections of analog components is to correct their parameters directly. For instance, if the value of a resistor or capacitor influences a time constant which has to be well controlled in a filter [20], it is possible to tune the value of the devices. Another example is the tuning of capacitors in DACs [21].

Figures 26 and 27 present the implementation of digitally adjustable components. By using the switches, it is possible to trim the value of the component. The sub-components can be connected in parallel (figure 26) or in series (figure 27).

The choice of the values of the components in the array can be optimized. Usually, such an array replaces a single component for which the required accuracy cannot be achieved without calibration.

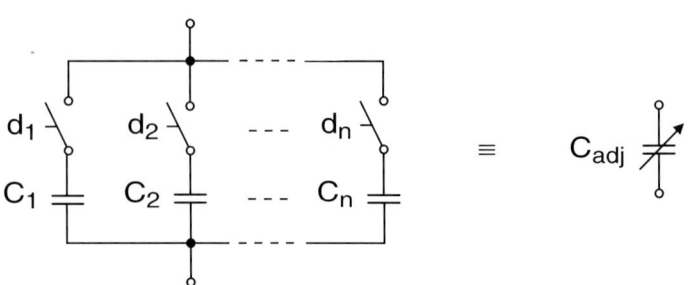

Figure 26. Parallel capacitor array

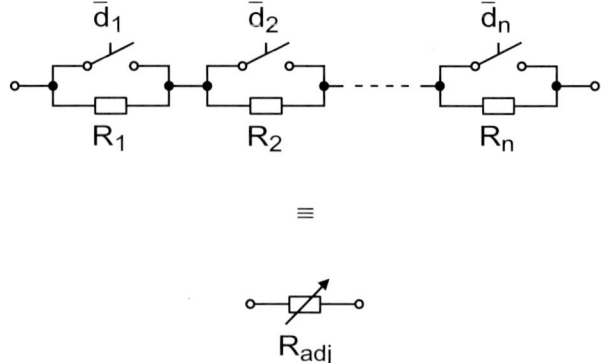

Figure 27. Series resistor array

5.1 Sizing

The main issue when designing component arrays is to find the weights of the different elements [22]. In all cases, the layout of the ladder should be done with care in order to match the constituent components and so that the variation between the characteristics of the elements is predicted by equation 2.1.

To keep consistent with the notation used previously and since the unit of the values of the elements in the array can correspond to any physical quantity, depending on the nature of the components (resistors, capacitors, ...), the value of component i is named b_i.

Chapter 3: Digital compensation circuits

To respect equation 3.5 even with mismatched components, b_i should be sized taking into account the component variation. In [21], a unilateral variation is supposed. Here, a bilateral tolerance function is used. Let's name δ the worst case mismatch between components, defined as:

$$\delta = \frac{b_{max} - b_{min}}{2(b_{max} + b_{min})} \qquad (3.14)$$

where b_{min} and b_{max} are respectively the minimum and maximum possible values of the component parameter b, taking into account the symmetrical bilateral deviation δ from the nominal value b_{nom}.

Another equivalent formulation of 3.14 is $b \in [b_{min}, b_{max}]$, with:

$$b_{min} = b_{nom}(1 - \delta), \, b_{max} = b_{nom}(1 + \delta) \qquad (3.15)$$

The parameter δ is derived from the matching parameters σ of the fabrication process. To guarantee an acceptable yield, the value $\delta = 3\sigma$ is usually chosen, so that more than 99.7% of the samples fall within the limits. However, this approach implies that all the components in the array are matched together as described in chapter 2, section 2. If this is not the case, δ can still represent component tolerances, but not mismatch in the strict sense.

Using this parameter δ, one can rewrite condition 3.5 to account for component mismatch in the component sizing:

$$(1 + \delta)b_i \leq (1 - \delta)\left(b_1 + \sum_{j=1}^{i-1} b_j\right) \qquad \forall i \in [2...n] \qquad (3.16)$$

In equation 3.16, the left term is the highest possible value of component b_i. The right part of the inequality represents the lowest possible sum of the less significant components (than i). Remember that if δ is extracted from matching parameters, this is valid only if all the components in the array are matched, so that the elements 1, ..., i-1 can be considered as one single equivalent component with the same matching properties.

If the values b_i are chosen at the limit of inequation 3.16, the radix of the ladder is:

$$r_i = \frac{b_i}{b_{i-1}} = \begin{cases} \frac{2(1-\delta)}{1+\delta} & i = 2 \\ \frac{2}{1+\delta} & i \in [3...n] \end{cases} \qquad (3.17)$$

It is noteworthy that this maximum allowable radix to guarantee correct operation depends directly on the intrinsic mismatch of the components. The less precise the components are, the lower the radix has to be and the more the converter is sub-binary. The optimal case with a radix-2 converter is realizable only with perfect components ($\delta = 0$).

Finally, one should note that in order to respect the matching rules (chapter 2, section 2.1), all the components should be made integer multiples of the least significant one (b_1). This implies that equation 3.17 is only theoretical and represents the best case, which is never implemented in practice.

6   CURRENT SOURCES

Compensating directly the imperfections of the components by tuning them is not always the best solution. Instead of compensating the source of imperfections, it is also possible to act on their consequences. For instance, the offset of an operational amplifier can be tuned by injecting a compensation current into an internal node of the circuit (see chapter 4).

This compensation current injection method is applicable to a large variety of situations. It enables the construction of generic compensation blocks which can be reused in different situations, and a systematic approach for inserting these trimming circuits in the systems to be calibrated.

This chapter presents two different structures of current-mode digital-to-analog converters. In this section, an implementation based on current mirrors is detailed. The rest of the chapter presents another DAC topology based on modified R/2R ladders.

6.1 Current-mirror DAC

Figure 28 shows the structure of the digital-to-analog converter [23]. The numbers near the transistors represent their respective W/L aspect ratios.

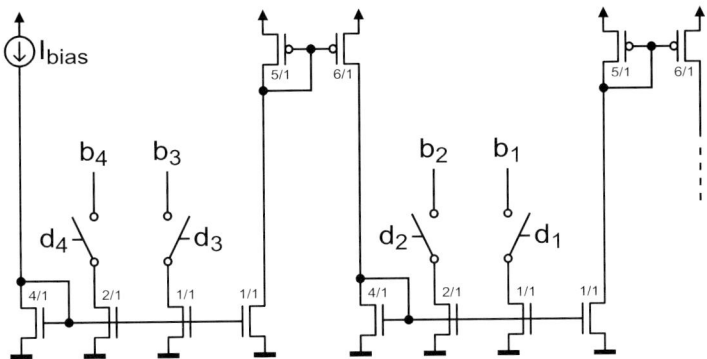

Figure 28. Sub-binary DAC based on current-mirrors

In each stage, the multiple output current mirror (bottom) is designed to divide the current into 2 binary-weighted output currents. The inter-stage current mirrors (top), combined with the input transistor of the next stage, perform a sub-binary current division, with a radix approximately equal to 1.7. Table 4 shows the current division performed by the circuit.

Table 4. Bit current values in the sub-binary DAC

bit (i)	current (b_i)
4	$\frac{1}{2}I_{bias}$
3	$\frac{1}{4}I_{bias}$
2	$\frac{3}{20}I_{bias}$
1	$\frac{3}{40}I_{bias}$

In this example, both stages are designed using the same transistor sizes. However, other stages can be are added as suggested on the right hand of the figure. Different transistor W/L ratios can become necessary to adapt the saturation voltages to the current flowing through the stage.

Since the array is sub-binary, the area of the circuit can obviously be made small. In [23], a 40 mil^2 area is achieved for a 12-bits DAC in a 2μm technology.

7 R/2R LADDERS

R/2R ladders are a well-known realization of binary-radix digital-to-analog converters. This section briefly introduces R/2R networks. Then, an implementation of R/2R ladders using MOS transistors as pseudo-resistors is studied in sections 8 and 9. Finally, section 10 presents a sub-binary radix version of this ladder, which can be used efficiently for successive approximations.

There are two variants of R/2R DACs. The first one is the voltage-mode network [24], which adds up fractions of a reference voltage to produce a voltage output. The second is the current-mode ladder, which collects split components of a reference current to produce a current output.

Figure 29 presents a 4 bits current-mode R/2R ladder. It is a regular arrangement of two values of resistors: R between the stages and 2R for each stage and as terminator. In each stage, the current is switched either to the output I_{out} or to a garbage current collector, in this case the ground. The switches are operated by the digital input word D of the DAC. The voltage in the current output node has to be equal to the voltage in the garbage collector, so that switching b_i has no effect on the current division in the ladder.

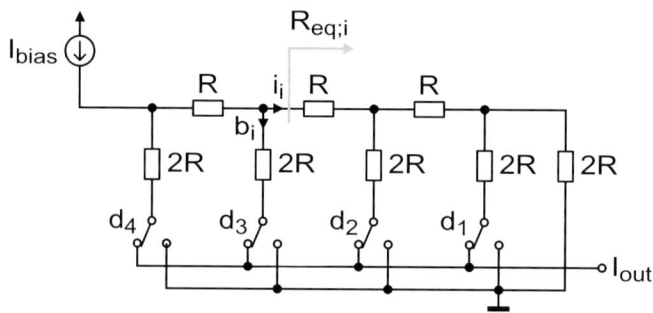

Figure 29. Current-mode R/2R ladder

In each stage of the network, the incoming current i_{i+1} is divided into 2 equal currents b_i and i_i:

Chapter 3: Digital compensation circuits

$$i_{i+1} = b_i + i_i = 2b_i \qquad (3.18)$$

The reason for this symmetrical division is that:

$$R_{eq;i} = 2R \qquad \forall i \in [1...n] \qquad (3.19)$$

Equation 3.19 can be proven by induction using the following property:

$$R_{eq;i+1} = R + (2R \parallel R_{eq;i}) \qquad (3.20)$$

The R/2R structure is generalized in [25], where a complete family of binary current dividers is analyzed and compared.

Concerning the implementation of R/2R converters, it is usually difficult to realize precise resistors with low area in modern fabrication processes. Another more practical solution is to use MOS transistors instead of resistors. This approach is detailed in the next sections.

8 LINEAR CURRENT DIVISION USING MOS TRANSISTORS

8.1 Principle

A convenient technique to implement the current dividers with MOS transistors instead of resistors is described in [26] and [27]. The principle is based on the symmetrical behavior of MOS transistors and the particular structure of their drain current equation:

$$I_D = \frac{W}{L} \int_{V_S}^{V_D} f(V_G, V) dV \qquad (3.21)$$

Where I_D is the drain current of the transistor, W and L its width and length respectively, and V_S and V_D the source and drain voltages. $f(V_G, V)$ is a function depending only on the gate voltage V_G and the voltage through the channel V, which is the integration variable. It is simply the inversion charge density in function of the channel voltage [28]:

$$f(V_G, V) = -\mu Q_i \qquad (3.22)$$

where μ is the electron/hole mobility. The equation for Q_i depends on the inversion mode of the transistor. In strong inversion (the inversion factor I_F is much larger than 1):

$$-Q_i \cong C_{ox} n (V_P - V) \qquad (3.23)$$

where C_{ox} is the gate oxide capacitance (per unit area), n the slope factor, and V_P the pinch-off voltage defined by:

$$V_P \cong \frac{V_G - V_{T0}}{n} \qquad (3.24)$$

V_{T0} is the threshold voltage. In weak inversion, (I_F much smaller than 1):

$$-Q_i \cong C_{ox}(n-1) U_T e^{\frac{\psi_0 - 2\phi_F}{U_T}} e^{\frac{V_P - V}{U_T}} \qquad (3.25)$$

where U_T is the thermodynamic potential, ψ_0 the surface potential and ϕ_F the bulk Fermi potential.

Figure 30 shows a graphical representation of the function $f(V_G, V) = -Q_i$. From the plot, I_D can be calculated as the shaded area A, corresponding to the integral of $f(V_G, V)$ between the source and drain voltages, multiplied by the aspect ratio W/L:

$$I_D = \mu \frac{W}{L} \int_{V_S}^{V_D} -Q_i \, dV = \mu \frac{W}{L} A \qquad (3.26)$$

As suggested by equation 3.21, since the function $f(V_G, V)$ is fixed by technological parameters, the drain current only depends on the source and drain voltages, and on the transistor aspect ratio W/L. This property allows to implement linear current division circuits using only MOS transistors, which act as pseudo-resistors. Although their characteristics are not linear, they can perform a linear current division, which is explained here on the basis of figure 31.

Chapter 3: Digital compensation circuits

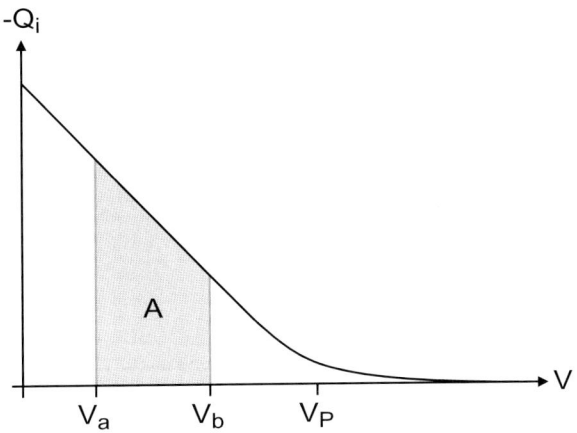

Figure 30. Normalized drain current of the MOS transistor

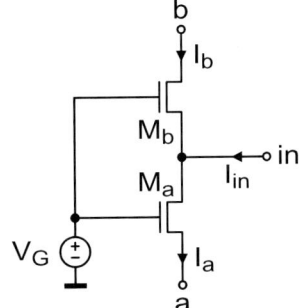

Figure 31. Current division circuit

In this circuit, both transistors have the same gate voltage V_G with respect to the substrate. Usually, they share the same bulk or well. The drain of M_a is connected to the source of M_b. This implies that in the graphical representation of figure 32, the areas A and B corresponding to the normalized (by W/L) currents I_a and I_b abut on to each other. In this initial situation, it is supposed that there is no input current ($I_{in} = 0$). The corresponding voltage at the input node is V_{in0}, the source voltage of M_a is V_a and the drain voltage of M_b is V_b.

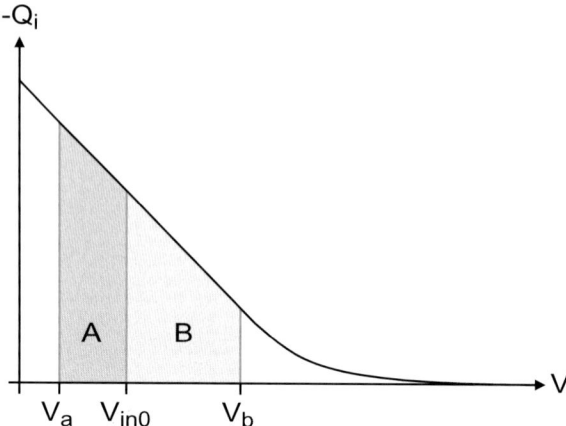

Figure 32. Current division without input current

If there is an input current ($I_{in} \neq 0$), it is divided and flows through M_a and M_b. If the new voltage at the input node is V_{in}, the difference ΔA between the initial areas and the present areas is represented in black in figure 33.

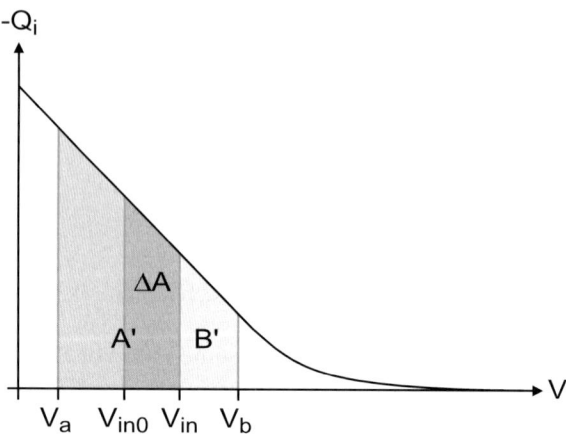

Figure 33. Current division with input

It is noteworthy that this differential area is the same for both M_a and M_b. In fact, one has:

$$A' = A + \Delta A \tag{3.27}$$

and

Chapter 3: Digital compensation circuits

$$B' = B - \Delta A \qquad (3.28)$$

Using equation 3.26, it is possible to calculate the ratio between the current flowing through M_a and the current flowing through M_b:

$$\frac{\Delta I_a}{\Delta I_b} = \frac{I'_a - I_a}{I'_b - I_a} = \frac{\frac{W_a}{L_a}(A' - A)}{\frac{W_b}{L_b}(B' - B)} = -\frac{W_a/L_a}{W_b/L_b} \qquad (3.29)$$

This current depends only on the ratio of the aspect ratios W_a/L_a and W_b/L_b of the two transistors performing the current division. Since the aspect of the function $f(V_G, V)$ is not important, the current is always divided linearly according to equation 3.29. The fact that the transistors are in the linear region or saturated, in weak or strong inversion does not influence the current division.

8.2   Second-order effects

The current division principle relies on the fact that both transistors share the same function $f(V_G, V)$. However, if the intrinsic parameters (V_{T0} for instance) of the transistors are not matched, this is no more the case and the current division becomes non-linear. The most disturbing effects are V_{T0} mismatches and channel length shortening [29]. To limit the effect of these imperfections, it is generally advised, if possible, to choose a large channel length, bias the transistors in the linear region and to use a high V_G.

8.3   Parallel configuration

Figure 34 shows two MOS transistors M_1 and M_2 connected in parallel, and the equivalent transistor M_{eq}.

Since M_1 and M_2 connected in parallel, one has:

$$\begin{aligned} V_{S;1} &= V_{S;2} = V_{S;eq} \\ V_{D;1} &= V_{D;2} = V_{D;eq} \end{aligned} \qquad (3.30)$$

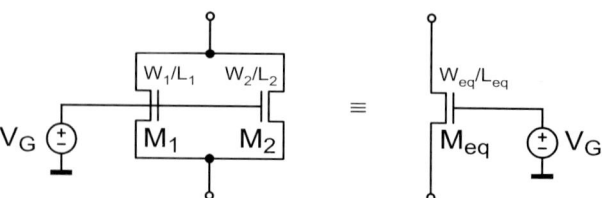

Figure 34. Equivalent transistor of two transistors in parallel

and

$$I_{D;eq} = I_{D;1} + I_{D;2} \qquad (3.31)$$

Using these properties, one can write, using equation 3.21:

$$\begin{aligned} I_{D;1} + I_{D;2} &= \frac{W_1}{L_1}\int_{V_{S;1}}^{V_{D;1}} f(V_G, V)dV + \frac{W_2}{L_2}\int_{V_{S;2}}^{V_{D;2}} f(V_G, V)dV \\ &= \left(\frac{W_1}{L_1} + \frac{W_2}{L_2}\right)\int_{V_{S;eq}}^{V_{D;eq}} f(V_G, V)dV \\ &= \frac{W_{eq}}{L_{eq}}\int_{V_{S;eq}}^{V_{D;eq}} f(V_G, V)dV \\ &= I_{D;eq} \end{aligned} \qquad (3.32)$$

From equation 3.32:

$$\frac{W_{eq}}{L_{eq}} = \frac{W_1}{L_1} + \frac{W_2}{L_2} \qquad (3.33)$$

The equivalent aspect ratio of two transistors in parallel is the sum of their respective aspect ratios. For instance, connecting two transistors with dimensions W/L in parallel is equivalent to one single transistor with dimensions 2W/L.

8.4   Series configuration

Figure 35 shows two MOS transistors M_1 and M_2 connected in series, and the equivalent transistor M_{eq}.

Chapter 3: Digital compensation circuits

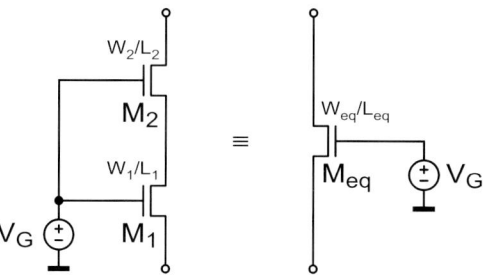

Figure 35. Equivalent transistor of two transistors in series

Since transistors M_1 and M_2 are connected in series:

$$V_{D;1} = V_{S;2} \qquad (3.34)$$

and

$$I_{D;eq} = I_{D;1} = I_{D;2} \qquad (3.35)$$

Using equation 3.34, one can write:

$$\int_{V_{S;eq}}^{V_{D;eq}} f(V_G, V) dV = \int_{V_{S;1}}^{V_{D;1}} f(V_G, V) dV + \int_{V_{S;2}}^{V_{D;2}} f(V_G, V) dV \qquad (3.36)$$

Simplifying equation 3.36 using 3.21 leads to:

$$\frac{L_{eq}}{W_{eq}} I_{D;eq} = \frac{L_1}{W_1} I_{D;1} + \frac{L_2}{W_2} I_{D;2} \qquad (3.37)$$

which can further be reduced because of equation 3.35:

$$\frac{L_{eq}}{W_{eq}} = \frac{L_1}{W_1} + \frac{L_2}{W_2} \qquad (3.38)$$

The inverse of the equivalent aspect ratio of two transistors in series is the sum of their respective inverse aspect ratios. For instance, connecting two transistors with dimensions W/L in series is equivalent to one single transistor with dimensions W/2L.

9 M/2M LADDERS

Using the properties presented in section 8, one can design an R/2R ladder using MOS pseudo-resistors instead of resistors. This network is named M/2M.

9.1 Principle

Figure 36 presents a 4 bits M/2M ladder.

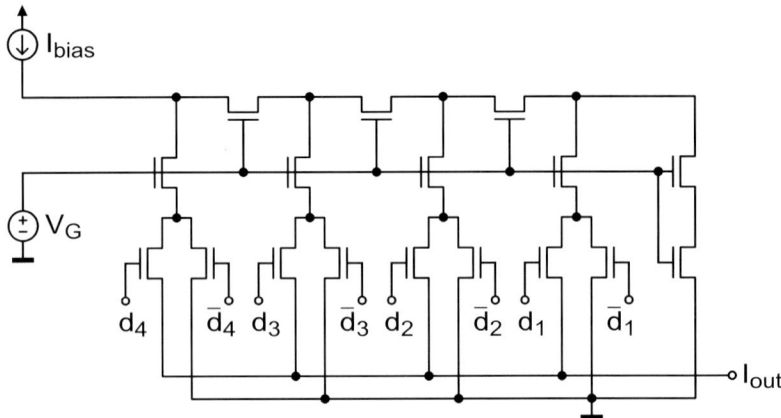

Figure 36. M/2M ladder

This M/2M ladder corresponds exactly to the R/2R network of figure 29. Each transistor acts as a single pseudo-resistor having a value equivalent to R, except the bottom-most transistors which act either as an open switch or also as R. In this way, it is possible to save one extra transistor since the switches are integrated in the pseudo R/2R ladder.

To make each transistor behave as a pseudo-resistor R, the rules presented in section 8 are applied. All the transistors in the ladder have the same dimensions W and L, and their gate voltage is equal to V_G. For the bottom-most transistors, the gate voltage has two possible values: If the switch has to be open, a gate voltage ensuring the blocking of the transistor is applied. When the switch is closed, it has to act as a pseudo-resistor with the same value as the other transistors in the ladder because the voltage V_G is applied to the gate.

If one chooses V_G = VDD, it is possible to use directly logic levels to drive the gates of the switch transistors. A high logic level (gate voltage equal to VDD) makes the corresponding transistor act as a pseudo-resistor, whereas a low logic level (gate voltage equal to 0) forces it to act as an open switch.

Chapter 3: Digital compensation circuits

This simplification is also possible in the case of a PMOS transistor ladder (figure 37), but using $V_G = 0$ and complementary logic levels.

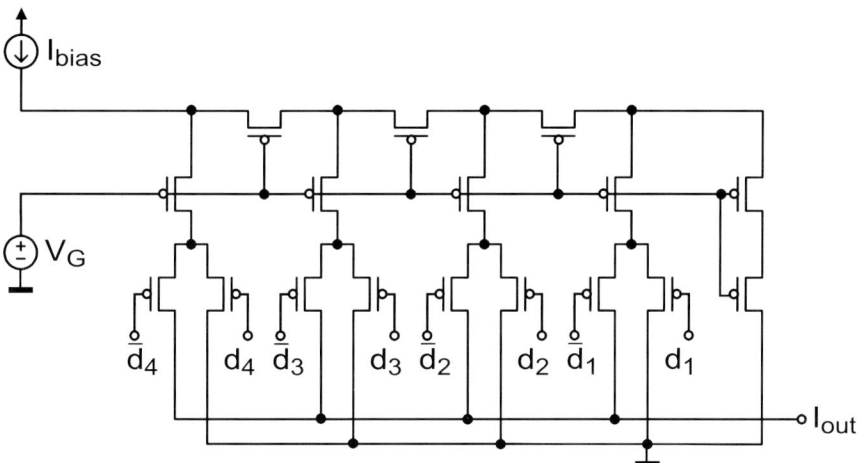

Figure 37. PMOS M/2M ladder

Finally, note that as for the R/2R ladder, the voltages in both current-collecting nodes (here I_{out} and the ground) have to be equal in order to perform the current division according only to the values of the pseudo-resistors.

9.2 Complementary ladder

It is also possible to design a complementary M/2M ladder, i.e. to use two MOS transistors in parallel as unit-value pseudo-resistor and one single transistor as a 2-unit value pseudo-resistor. Figure 38 presents the resulting network.

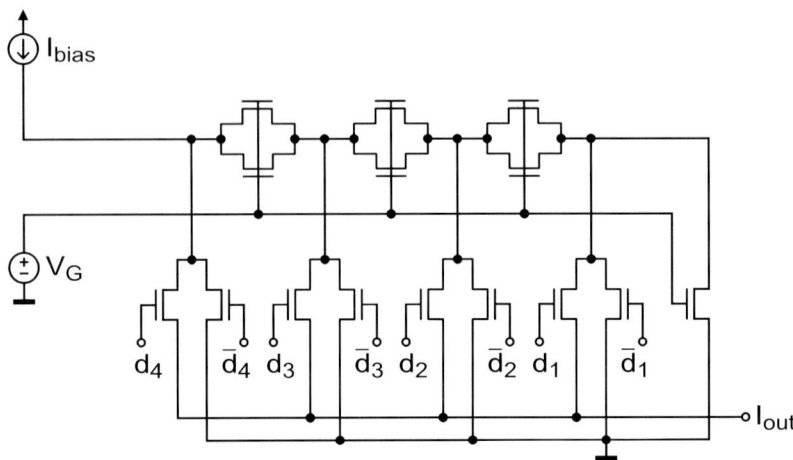

Figure 38. Inverse M/2M ladder

The advantage of this structure over the conventional one presented in section 9.1 is a doubled equivalent aspect ratio (W/L instead of W/2L). This implies that for equivalent transistor dimensions and an equal bias current I_{bias}, the complementary ladder causes a smaller voltage drop between the bias current input and the current collector nodes. This can be advantageous if the available voltage range is small.

9.3 Second-order effects

The M/2M ladder can be used to realize digital-to-analog converters. However, the second-order effects discussed in section 8.2 impact the linearity and limit the achievable resolution. The best result published by today is a 10 bits resolution M/2M converter [29], in a 1μm process with a 5V power supply. In this design, the transistors composing the ladder have an aspect ratio of W/L = 30μm/30μm and an important matching effort is made. Furthermore, the ladder being sensitive to a voltage difference between the two current collector nodes, a regulator circuit is implemented.

Other analyses of the impact of the mismatch of the transistors in a M/2M ladder on the best-achievable resolution of digital-to-analog converters can be found in [30], [31] and [32]. They all emphasize the need for component matching and high-area devices to achieve high resolutions.

9.4 Trimming

Another solution to increase the resolution of M/2M ladders is to use calibration. A moderate effort is made to design a ladder with an acceptable resolution and a calibration is done to increase it. An algorithm capable of calibrating R/2R ladders is proposed in [33].

In [34], the M/2M ladder is calibrated by adjusting the threshold voltage V_T of the transistors. This is done by placing some of the transistors into separate wells and trimming their bulk voltage to change V_T. This approach is rather difficult to implement since it requires separate wells, which are not always available, and which impact negatively on the matching accuracy of the transistors. Furthermore, it needs a separate DAC for each transistor which is trimmed, or a DAC and a multiplexer associated with individual memories for each trimmed device.

A similar result is achieved by slightly modifying the gate voltage of some transistors of the ladder instead of changing their V_T. This approach is implemented [35] in a current divider circuit. The technique can be extended to M/2M structures, but it has the same drawbacks as the V_T adjustment.

10 R/XR LADDERS

A high-precision DAC is not necessary to perform successive approximations. With only a slight modification, it is though possible to make the ladder sub-binary and to relax the design constraints.

10.1 Principle

Figure 39 presents a 4-stages R/xR ladder, which is a generalization of the R/2R ladder.

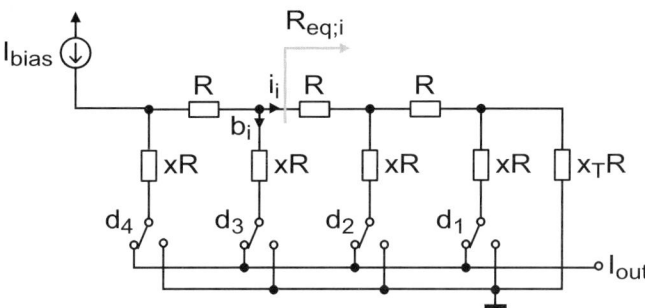

Figure 39. R/xR ladder

The resistor in the vertical branches has here a value xR, and the terminator $x_T R$. The case with

$$x = x_T = 2 \qquad (3.39)$$

corresponds to the R/2R ladder. On the opposite, choosing

$$x > 2 \qquad (3.40)$$

makes the converter sub-binary. This special case is analyzed below. A theoretical study of the R/xR structure can be found in [36]. Here, a more circuit-oriented analysis is done.

In each stage of the network, the incoming current i_{i+1} is divided into 2 currents b_i and i_i:

$$i_{i+1} = b_i + i_i \qquad (3.41)$$

with

$$b_i = i_{i+1} \frac{R_{eq;i}}{xR + R_{eq;i}} \qquad (3.42)$$

and

$$i_i = i_{i+1} \frac{xR}{xR + R_{eq;i}} \qquad (3.43)$$

More interesting is the ratio ρ_i between these two currents:

$$\rho_i = \frac{b_i}{i_i} = \frac{R_{eq;i}}{xR} \qquad (3.44)$$

Chapter 3: Digital compensation circuits

In fact, this ratio can be used to derive the condition which allows the ladder to be used for successive approximations. Let's make two small modifications in the ladder: Using the terminator current as a bit in the converter instead of dumping it, and numbering the bits by taking care to exchange the last two currents to have:

$$b_1 = i_2 = i_3 \frac{x_T}{x_T + x} \qquad (3.45)$$

and

$$b_2 = i_3 \frac{x}{x_T + x} \qquad (3.46)$$

Figure 40 shows the resulting circuit.

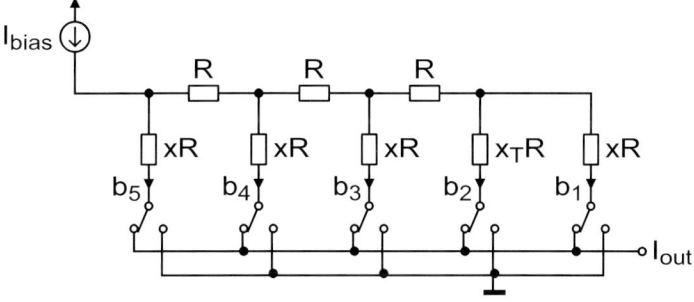

Figure 40. Modified R/xR ladder

10.2 Working condition

In each stage of the ladder in figure 40, the remaining current i_i is equal to the sum of the less significant bits:

$$i_i = \sum_{j=1}^{i-1} b_j \qquad (3.47)$$

To allow the converter to work with a successive approximations algorithm, it is sufficient that in all stages but the last one the remaining current (which is also the sum of all less significant bits), is higher than the corresponding bit current. In other words, the value of the equivalent resistance of the remaining right part of the ladder has to be lower than the vertical resistor xR. By imposing:

$$\rho_i \leq 1 \qquad \forall i \in [3...n] \qquad (3.48)$$

which is, by equation 3.44, equivalent to:

$$b_i \leq i_i \qquad \forall i \in [3...n] \qquad (3.49)$$

the working condition of the successive approximations algorithm (equation 3.5) is respected indeed, because:

$$b_i \leq \sum_{j=1}^{i-1} b_j \leq b_1 + \sum_{j=1}^{i-1} b_j \qquad \forall i \in [3...n] \qquad (3.50)$$

And for i = 2, in order to guarantee the working condition, one must have:

$$b_2 \leq 2b_1 \Rightarrow x \leq 2x_T \qquad (3.51)$$

10.3 Terminator calculation

One particularly interesting value for x_T is the one that ensures that:

$$R_{eq;3} = R + (x_T R \parallel xR) = R + \frac{R^2 x_T x}{R(x_T + x)} = x_T R \qquad (3.52)$$

because in this case, all the $R_{eq;i}$ (i > 2) are equal to $x_T R$. Simplifying equation 3.52 gives:

$$x_T^2 - x_T - x = 0 \qquad (3.53)$$

The only solution of this equation is:

Chapter 3: Digital compensation circuits

$$x_{T;stat} = \frac{1+\sqrt{1+4x}}{2} \qquad (3.54)$$

Choosing $x_T = x_{T;stat}$ makes the ladder stationary, i.e. in all the stages of the ladder, the equivalent resistor $R_{eq;i}$ (i > 2) has the same value $x_{T;stat}R$ and the current is divided exactly with the same ratio ρ_{stat}:

$$\rho_{stat} = \frac{x_{T;stat}R}{xR} = \frac{1+\sqrt{1+4x}}{2x} \qquad (3.55)$$

In the last stage, the ratio is inverted. It is now obvious that the two last bits had to be exchanged to ensure that $b_2 \geq b_1$. This is because:

$$x_{T;stat}R \leq xR \qquad \forall x \geq 2 \qquad (3.56)$$

This is the case indeed (see equations 3.40 and 3.54). Furthermore, to ensure proper functioning of the last stage of the ladder, equation 3.51 has to be respected. It is easy to prove by replacing 3.54 in 3.51 that this is the case if:

$$x \leq 6 \qquad (3.57)$$

Such a high value of x is never implemented anyway. It leads to a very low radix that makes the converter inefficient.

10.4 Terminator implementation

As shown by equation 3.54, the ideal value of the terminator is not an integer. To solve the implementation problem, [36] proposed to realize the terminator using additional R/xR stages. But this is not necessary, because an integer value of x_T satisfying all the working conditions can be chosen. In fact, the sufficient condition for x_T is simply:

$$\frac{x}{2} \leq x_T \leq x_{T;stat} \qquad (3.58)$$

The left inequation is directly derived from equation 3.51, whereas the right part is explained as follows: To ensure the working condition, section 10.2 shows that a sufficient condition is:

$$R_{eq;i} \leq xR \qquad (3.59)$$

This is ensured in the stationary case (equation 3.56). If the value of x_T is voluntarily chosen smaller than $x_{T;stat}$, according to equation 3.58, it can be proven that the $R_{eq;i}$ become a series converging to the value $x_{T;stat}R$. Furthermore, the series has the property of being strictly increasing for increasing values of i. This is due to the fact that the first value is chosen lower than the limit value $x_{T;stat}R$.

Figure 41 shows the effect of choosing a 2R terminator ($x_T = 2$) in a R/3R ladder (x = 3). The value of the equivalent resistor at each stage is plotted. The equivalent value converges quickly to the limit value ($x_{T;stat} \cong 2.3$ with x = 3), without ever exceeding it. The limit $x_{T;stat}$ is plotted in grey in the figure.

To explain the result presented in this section in other words, choosing $x_T \leq x_{T;stat}$ puts the converter in an even more favorable case with respect to the working condition. Indeed, the highest $R_{eq;i}$ value is reached in the last stage (MSB) instead of being equal in every stage to $x_{T;stat}R$ (sationary case). This is particularly important when the ladder is sized for a given component matching quality. The liberty given for the choice of the terminator value can be exploited to simplify the ladder implementation, without impacting negatively on the immunity of the circuit to variations of the component values.

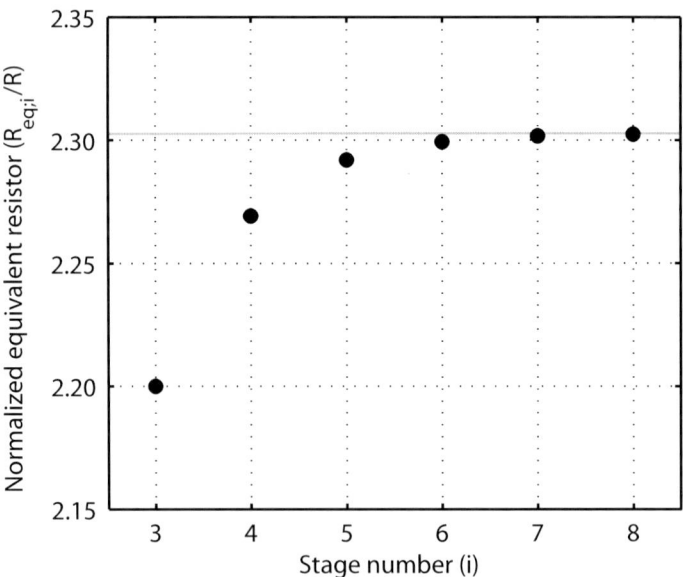

Figure 41. 2R terminator in a R/3R ladder

10.5 Ladder sizing

The same analysis as in section 5.1 is made here for the R/xR converter to calculate x as a function of the component matching parameter δ. The terminator sizing is covered in the next section.

Remember that if all the components composing the converter are not matched using the rules of chapter 2, section 2.1, δ represents the tolerance of the components.

Let's rewrite equation 3.49 in the worst case:

$$b_i(1+\delta) \leq i_i(1-\delta) \tag{3.60}$$

Which is, using equation 3.44, equivalent to:

$$R_{eq;i}(1+\delta) \leq xR(1-\delta) \tag{3.61}$$

The worst case (highest possible value) for $R_{eq;i}$ is $x_{T;stat}R$ if the terminator is chosen according to equation 3.58:

$$x_{T;stat}R(1+\delta) \leq xR(1-\delta) \tag{3.62}$$

From equation 3.53, one can write:

$$x = x_{T;stat}(x_{T;stat} - 1) \tag{3.63}$$

Replacing 3.63 in 3.62 and simplifying:

$$(1+\delta) \leq (x_{T;stat} - 1)(1-\delta) \tag{3.64}$$

Finally, using 3.54 and simplifying again, one has:

$$x \geq 2\frac{1+\delta}{(1-\delta)^2} \tag{3.65}$$

The optimum resistor ratio can thus be determined, for a given imperfection level δ, using equation 3.65. This relation can also be reversed, in order to calculate the allowable error δ in function of the resistor ratio x:

$$\delta \leq 1 + \frac{1-\sqrt{1+4x}}{x} \tag{3.66}$$

10.6 Terminator sizing

For the terminator, the left inequation 3.58 can also be rewritten in the worst case:

$$\frac{x}{2}(1+\delta) \leq x_T(1-\delta) \qquad (3.67)$$

Which is equivalent to:

$$x_T \geq x\frac{1+\delta}{2(1-\delta)} \qquad (3.68)$$

In the worst case, the right side of inequation 3.58 remains:

$$x_T \leq x_{T;stat} \qquad (3.69)$$

because equation 3.62 already accounts for the tolerance $(1+\delta)$ on x_T. Combining equations 3.68 and 3.69 results in:

$$x\frac{1+\delta}{2(1-\delta)} \leq x_T \leq \frac{1+\sqrt{1+4x}}{2} \qquad (3.70)$$

If the left inequality is violated, i.e. x_T is chosen too small, the local radix r_2 (the b_2/b_1 ratio) becomes higher than 2. In this case, the converter works perfectly, but with a LSB corresponding to b_2 instead of b_1. If the right inequality 3.70 is not respected, the converter works, but only if the tolerances of the components are sufficiently small, at any rate smaller than the limit value δ_{max} calculated from equation 3.66:

$$\delta_{max} = 1 + \frac{1-\sqrt{1+4x}}{x} \qquad (3.71)$$

The upper limit for x_T (equation 3.69) ensures that the worst case for the current division is reached in the highest stage of the converter. If on the contrary, x_T becomes larger than $x_{T;stat}$, the worst case is reached in the third stage, because the $R_{eq;i}$ series is then strictly decreasing. The working condition in the third stage is still $\rho_3 \leq 1$ (equation 3.48), which becomes in the worst case:

Chapter 3: Digital compensation circuits

$$\rho_3 = \frac{R_{eq;3}(1+\delta)}{xR(1-\delta)} = \frac{R + (xR \parallel x_T R)}{xR} \cdot \frac{(1+\delta)}{(1-\delta)} \quad (3.72)$$

$$= \frac{1 + \frac{xx_T}{x + x_T}}{x} \cdot \frac{1+\delta}{1-\delta} \leq 1$$

which gives:

$$\delta \leq \frac{x^2 - x - x_T}{x^2 + x + 2xx_T + x_T} \quad (3.73)$$

This equation has no solution when:

$$x_T > x^2 - x \quad (3.74)$$

as in this case, $R_{eq;3}$ becomes larger than xR and even with perfect components, ρ_3 is higher than 1. Figure 42 shows the maximum allowable mismatch δ in function of the terminator value x_T in the case of a R/3R ladder ($x = 3$). When $x_T \leq x_{T;stat}$, equation 3.66 applies, whereas equation 3.73 is valid when $x_T \geq x_{T;stat}$. For $x_T = x_{T;stat}$ both are equivalent, as expected. In the case of an R/3R ladder, $\delta_{max} \cong 0.13$ (equation 3.71) and $x_{T;stat} \cong 2.30$ (equation 3.54). The minimum value for x_T when $\delta = \delta_{max}$ is calculated with equation 3.68 and is about 1.95.

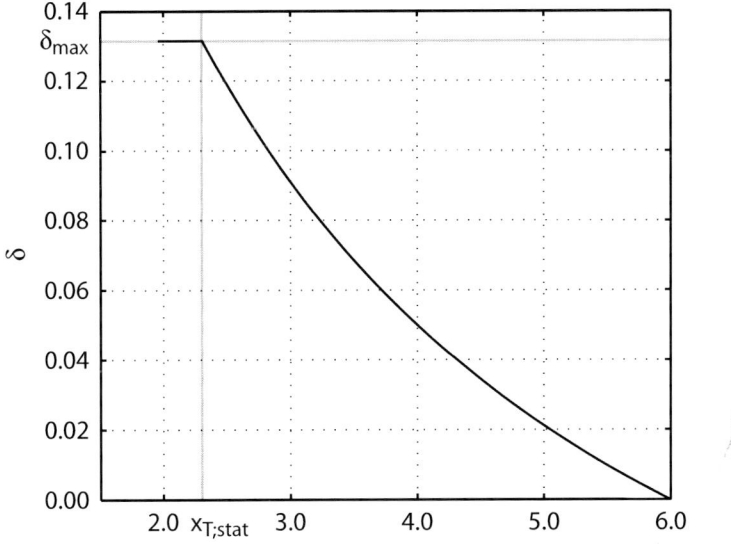

Figure 42. Maximum allowable mismatch in function of x_T

As shown here above, choosing $x_T > x_{T;stat}$ reduces the allowable mismatch and should thus be avoided. The optimum choice is $x_{T;stat}$ because it still allows $\delta = \delta_{max}$ without decreasing the local radix in the last stages. This happens when $x_T < x_{T;stat}$ because the equivalent resistor in the last stages is lower than $x_{T;stat}R$ (figure 41). This reduction also decreases ρ (equation 3.44) and thus the local radix, which depends on ρ (see equation 3.75). This choice is thus also sub-optimal as it diminishes the resolution of the converter. But if an integer value has to be chosen to simplify the realization of the terminator, it is better to choose a value $x_T \leq x_{T;stat}$.

The two last sections have shown how to size a R/xR ladder. Knowing the mismatch between components, it is possible to choose a resistor ratio x with equation 3.65, and a terminator with 3.70 that allows the use of the converter with a successive approximations algorithm. As is to be expected, these equations would speak in favor of the R/2R ladder if perfect components ($\delta = 0$) existed.

10.7 Radix

In the case where $x_T = x_{T;stat}$, the radix of the ladder (for $i > 2$) can be easily calculated:

$$R = \frac{b_i}{b_{i-1}} = \frac{b_i}{i_i}\left(1 + \frac{1}{\rho_{stat}}\right) = \rho_{stat}\left(1 + \frac{1}{\rho_{stat}}\right) = \rho_{stat} + 1 \qquad (3.75)$$

By taking the limit case of inequation 3.65 and replacing it in 3.55 gives:

$$\rho_{stat} = \frac{1-\delta}{1+\delta} \qquad (3.76)$$

And substituting the latter in 3.75:

$$R = \frac{2}{1+\delta} \qquad (3.77)$$

Chapter 3: Digital compensation circuits

Considering the last stage of the ladder, the limit choice for x_T is given by equation 3.68 and is:

$$r_2 = \frac{b_2}{b_1} = \frac{xR}{x_T R} = \frac{2(1-\delta)}{1+\delta} \tag{3.78}$$

Obviously, choosing this limit value for the terminator makes the ladder non-stationary and equation 3.77 not valid any more. However, equations 3.77 and 3.78 reflect the maximum allowable radixes as a function of the mismatches (or tolerances) of the components, respectively in the ladder and in the last stage.

A particularly remarkable fact is that these two equations are *exactly equivalent* to equation 3.17, which is the limit condition for component arrays. This is not surprising however, since even if the two structures are different, they both have exactly the same function and the same component mismatches. Achieving equal performance is a logical result, even if its proof is not trivial. To gain a little more insight in the relation of the radix with the mismatch, figure 43 plots the function of equation 3.77.

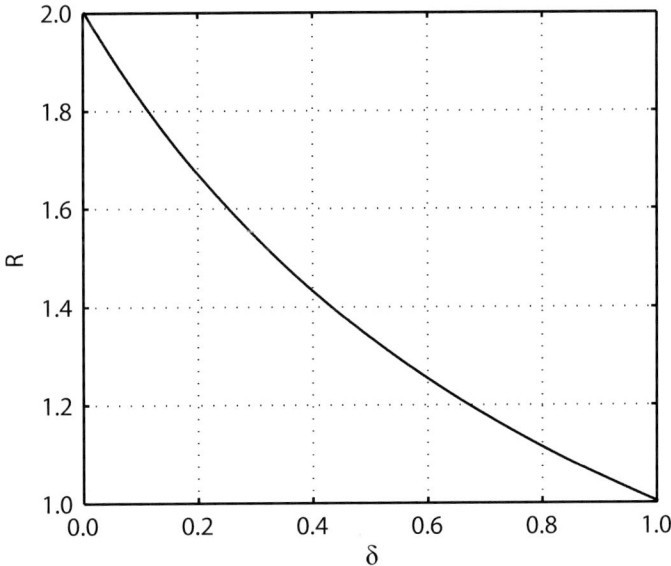

Figure 43. Best-achievable radix with a sub-binary converter

With perfect components, δ = 0 and the radix can be binary. This corresponds to a R/2R ladder or a binary-weighted component array.

With components having 100% mismatch in the worst case, the best achievable radix is 1, which corresponds to a thermometer coding (all the bits have the same nominal value).

Between these two extreme situations, the function is 1/x shaped. The largest loss in high radix value possibilities is made for a small δ increase when δ is close to 0. Any further increment has less impact, since the slope of the function decreases (in absolute value) when δ increases.

11 M/2⁺M LADDERS

Using the same technique as for M/2M ladders, some R/xR ladders can easily be realized using pseudo-resistors. Two particularly interesting variants of M/2⁺M (sub-binary) ladders are the M/3M and M/2.5M ladders. They are presented in this section.

11.1 M/3M ladders

The M/3M is the implementation of the R/3R structure using MOS transistors to realize pseudo-resistors of unit value R. Figure 44 presents an M/3M ladder.

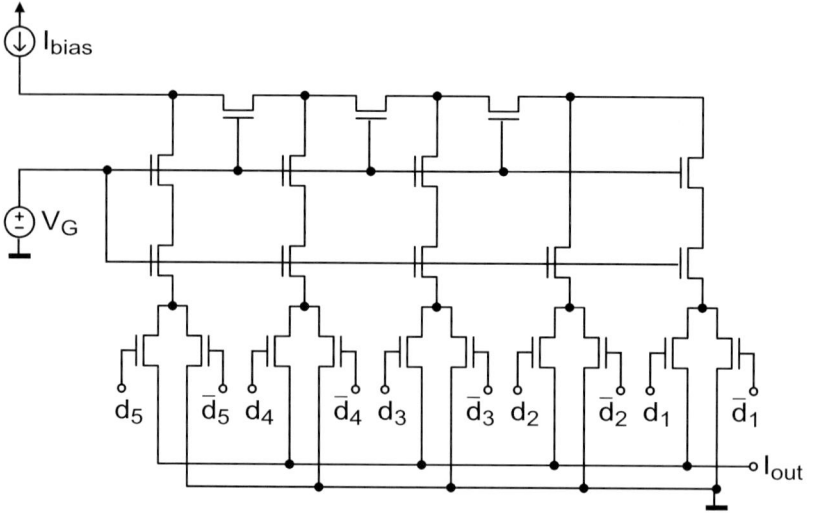

Figure 44. M/3M ladder

Chapter 3: Digital compensation circuits 63

In each stage, the horizontal R resistor is replaced by a single transistor, and the vertical 3R resistor is replaced by 3 transistors in series. The bottommost resistor is duplicated to also implement the switch, exactly as in the M/2M ladder (see section 9.1).

For the terminator, an equivalent 2R pseudo-resistor is used. This choice is done to simplify the implementation. In fact, the minimum and maximum values for x_T are calculated using equation 3.70 (δ calculated with equation 3.66). They are approximately equal to 1.95 and 2.30, respectively. The ideal value of 2.30 is rather difficult to realize using unity pseudo-resistors. On the other hand, the integer value 2, which lies in the acceptable interval, can be realized using only 2 transistors. The slight drawback is that the radix of the first stages of the ladder is in this case sub-optimal.

Table 5 presents the main characteristics of the M/3M (and R/3R) ladder. The equations used to calculate the values of the parameters are mentioned.

Table 5. Characteristics of the M/3M ladder

parameter	symbol	value	equation
maximum mismatch	δ_{max}	13 %	3.66
radix	R	1.77	3.77
current division ratio	ρ_{stat}	0.77	3.55
stationary terminator	$x_{T;stat}$	2.30	3.54
minimum terminator	$x_{T;min}$	1.95	3.70
maximum terminator	$x_{T;max}$	2.30	3.70

The maximum allowable mismatch (or tolerance) is $\delta_{max} \cong 0.13$ (13 %). If this condition is respected, the current division ratio which has a nominal value of 0.77 never exceeds 1.0. This ensures that the converter can be used with a successive approximations algorithm. The nominal radix is 1.77, which is significantly lower than 2. This means that to achieve a similar resolution, an M/3M converter needs more stages than a M/2M DAC. But the M/3M is much more robust to variations of the component values than its optimal counterpart, which makes it more attractive for applications with successive approximations.

11.2 M/2.5M ladders

The M/2.5M ladder is a compromise between the M/3M and the M/2M ladders. Figure 45 presents such a network.

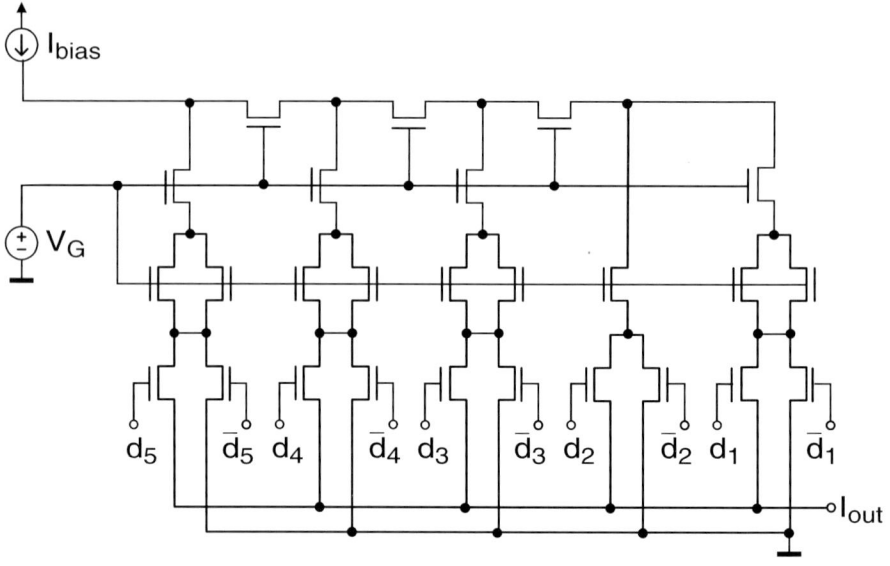

Figure 45. M/2.5M ladder

Chapter 3: Digital compensation circuits 65

The 2.5R pseudo-resistor (vertically) is realized using two unit value transistors (the upper-most one and the switch) in series with a parallel construction of two more unit transistors (in the middle). The terminator is still a 2R pseudo-resistor, for simplicity reasons. Note that this value is contained in the acceptable interval for x_T, which is [1.45, 2.16].

Table 6. Characteristics of the M/2.5M ladder

parameter	symbol	value	equation
maximum mismatch	δ_{max}	7.3 %	3.66
radix	R	1.86	3.77
current division ratio	ρ_{stat}	0.86	3.55
stationary terminator	$x_{T;stat}$	2.16	3.54
minimum terminator	$x_{T;min}$	1.45	3.70
maximum terminator	$x_{T;max}$	2.16	3.70

Table 6 presents the main characteristics of the M/2.5M (and R/2.5R) ladder. The maximum allowable mismatch (or tolerance) is $\delta_{max} \cong 7.3$ %. This implies that the components need to be more precisely matched to implement a M/2.5M ladder than a M/3M one. But on the other hand, the nominal radix is increased to 1.86 (instead of 1.77 for the M/3M). To achieve the same resolution, the M/2.5M ladder thus needs less stages, which is a clear advantage.

11.3 Ladder selection and other M/2^+M ladders

The choice of the ladder (M/2.5M or M/3M) is dictated by the mismatch or tolerance δ of the components. Equation 3.65 allows to calculate the minimum admissible value of x, which is the optimal value for a given δ. Figure 46 plots the function, and table 7 details the implementation of the xR resistor.

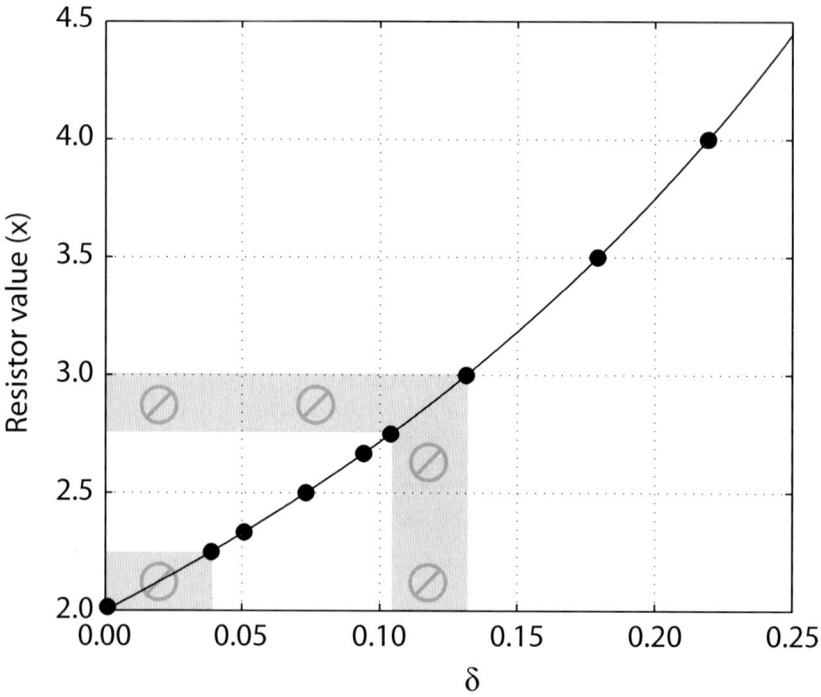

Figure 46. M/2⁺M ladder selection

The shaded areas in figure 46 correspond to values of δ and x (x ∈ [2, 3]) for which the resistor xR cannot be implemented with a small number of devices (less than 6). The total number of transistors per stage is calculated by adding 2 (one additional switch, plus the horizontal unity resistor) to the corresponding value in the right-most column of table 7.

Chapter 3: Digital compensation circuits

Table 7. 2^+ resistor implementation

δ_{max} [%]	Resistor value (x)	Implementation	Resistor #
0.0	2	R + R	2
3.9	2.250	R + R + (R || R || R || R)	6
5.1	2.333	R + R + (R || R || R)	5
7.3	2.5	R + R + (R || R)	4
9.4	2.666	R + R + (R || (R + R))	5
10.4	2.750	R + R + (R || (R + R + R))	6
13.1	3	R + R + R	3
17.9	3.5	R + R + R + (R || R)	5
21.9	4	R + R + R + R	4

The goal of M/2^+M converters is to allow the realization of *small* converters for successive approximations. Using large transistors to increase the matching (decrease δ), or using many transistors is contrary to this philosophy. Furthermore, the layout of the ladder (see section 11.6) usually does not respect all the matching rules of chapter 2, section 2.1. Consequently, increasing the size of the devices does not help much at any rate. From these findings, the following design guidelines should be applied:

1. Choose a unity device as small as possible, just ensuring that δ is not extremely high.
2. Estimate the tolerance δ.
3. Choose a simple ladder implementation using the figure and table in this section, preferably M/3M or M/2.5M.

The effectiveness of these guidelines is confirmed by measurements in section 11.7.

11.4 Current collector design

A possible design option for the current collector is to use current mirrors. The advantage is the simplicity of realization compared to other solutions like the operational amplifier trans-resistance circuit for example (figure 62).

Figure 47 shows a complete M/3M ladder with current collectors. In this case, a differential output is implemented.

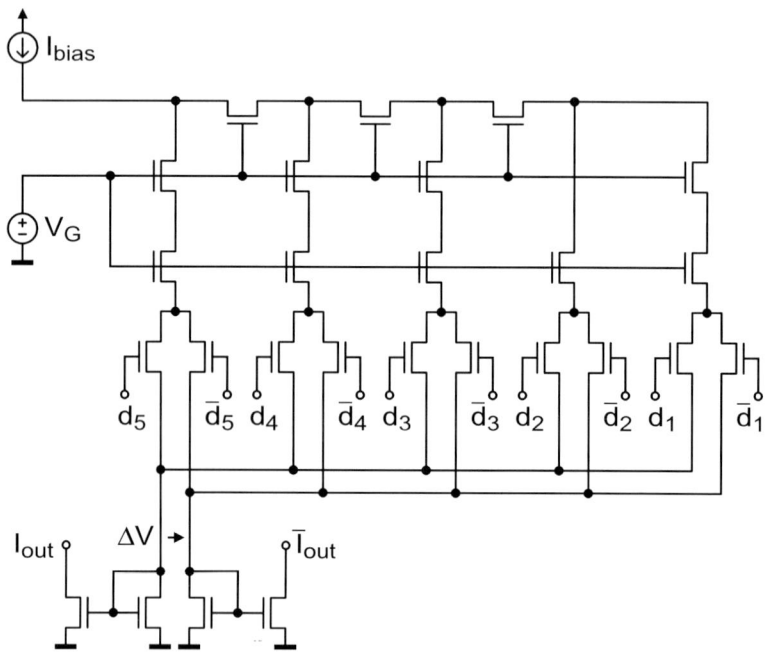

Figure 47. Current mirror as M/3M current collector

The main drawback of this approach is the differential voltage ΔV that builds up when the collected currents are different. This in turn causes an imbalance in the current division, which is not problematic for successive approximations as shown below.

Figure 48 shows an example of voltage/current characteristics of the current collector nodes, with two different aspect ratios.

Chapter 3: Digital compensation circuits

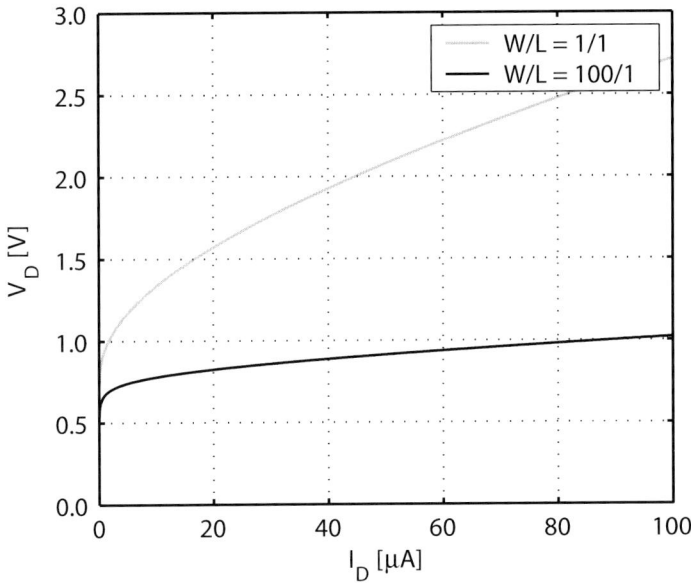

Figure 48. Voltage/current characteristics of a diode-connected transistor

The diode-connected transistors are always in saturation mode (because $V_D = V_G$), and the equation of their drain current is:

$$I_D = \frac{\mu C_{ox}}{2n} \cdot \frac{W}{L}(V_D - V_{T0})^2 \qquad (3.79)$$

where μ is the electron/hole mobility and C_{ox} the oxide capacitance.

The function is strictly increasing, and the W/L ratio determines the steepness of the slope above the threshold voltage. Increasing the W/L ratio diminishes the influence of a variation of the collected current on the voltage, and thus on the worst-case differential voltage ΔV. Furthermore, the absolute voltage drop is also reduced.

The effect of this voltage/current characteristics on the converter can be analyzed as follows: The working condition (equation 3.50) states that the sum of the remaining currents (less significant bits) should always be greater than the currently tested bit. If the ladder is correctly designed to respect this condition, it can be shown that the voltage drop caused by the current mirrors is not problematic. Suppose that the algorithm has already tested the most significant bits and is currently examining bit b_k. The current in the ladder can be divided into:

$$I_{kept} = \sum_{i=k+1}^{n} d_i b_i \qquad (3.80)$$

I_{kept} is the sum of the currents that have already been kept by the algorithm.

$$I_{rejected} = \sum_{i=k+1}^{n} \overline{d_i} b_i \qquad (3.81)$$

$I_{rejected}$ is the sum of the currents that have already been rejected by the algorithm.

$$I_{test} = b_k \qquad (3.82)$$

I_{test} is the current that is currently tested.

$$I_{remain} = \sum_{i=1}^{k-1} d_i b_i \qquad (3.83)$$

I_{remain} is the current that remains to be tested (the sum of the less significant bits). Equation 3.50 ensures that $I_{remain} \geq I_{test}$.

All these currents correspond to the ideal situation where the current collectors are perfect and ensure that $\Delta V = 0$ always. Let's consider what happens to these currents in the case where the current collectors are current mirrors.

At step k in the algorithm, the tested current I_{out} corresponds to $I_{kept} + I_{test}$, whereas $\bar{I}_{out}$ corresponds to $I_{rejected} + I_{remain}$. The term "corresponds" is chosen on purpose, since the currents are not summed exactly by the current collectors. This is due to the fact that increasing the current in one of the collectors also increases the corresponding voltage. This in turn causes a current reduction, since the ladder behaves roughly like a resistor. The ideal currents are not really summed up in the current collectors.

Figure 49 shows the situation before and after the decision of the successive approximations algorithm at step k on an example.

Chapter 3: Digital compensation circuits 71

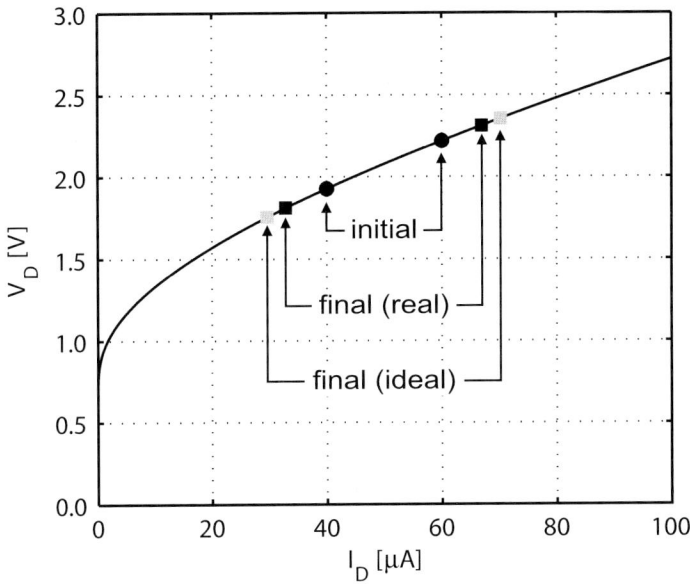

Figure 49. Successive approximations with current mirrors as collectors

The bias current of the ladder is 100 µA, I_{kept} = 50 µA, $I_{rejected}$ = 20 µA, I_{test} = 10 µA and I_{remain} = 20 µA. The solid points show the initial situation during the test: The sum of the currents of the previously kept bits and currently tested one is $I_{kept} + I_{test}$ = 60 µA, whereas the sum of the rejected and remaining current is $I_{rejected} + I_{remain}$ = 40 µA.

If the algorithm rejects the current bit k, the sum of the less significant bits must be higher than the just rejected bit. In other words, the output current corresponding to $I_{kept} + I_{remain}$ must be higher or equal to the previous one corresponding to $I_{kept} + I_{test}$. In the case of an ideal current collector, this situation corresponds to the grey squares. Since the voltages have been unbalanced however, the real situation is indicated by the black squares.

After the decision, the current I_{out} corresponds to $I_{kept} + I_{remain}$, whereas it was $I_{kept} + I_{test}$ before. It is thus increased because $I_{remain} \geq I_{test}$, and causes a voltage increase in the corresponding current collector. This in turn tends to somewhat diminish the current. On the other hand, the decrease of current in $\bar{I}_{out}$, which corresponds to $I_{rejected} + I_{test}$, provokes a voltage decrease which attenuates the reduction. This indeed corresponds to the black squares, which are located between the grey squares and the black points.

Although the current difference is not so high anymore as in the ideal case, the current corresponding to the end situation is still higher than at the beginning and ensures the correct functioning of the algorithm. The use of a current mirror as collector has no effect on the working condition.

11.5 Complementary ladders

As for M/2M ladders, the complementary versions of M/3M and M/2.5M ladders present the advantage of causing less voltage drop in the same conditions. They are built using the same procedure as for the complementary M/2M ladder (see section 9.2).

11.6 Layout

One advantage of M/xM structures is that their layout is simple. It is possible indeed to layout one stage of the ladder and to replicate it as many times as necessary (i.e. by the number of stages of the converter). It is even possible to fully automate the layout since the structure is regular. A typical placement of the transistors of one stage of a M/2.5M ladder is presented in figure 50. The sources and drains are drawn in light grey and the dark grey is used for the gates.

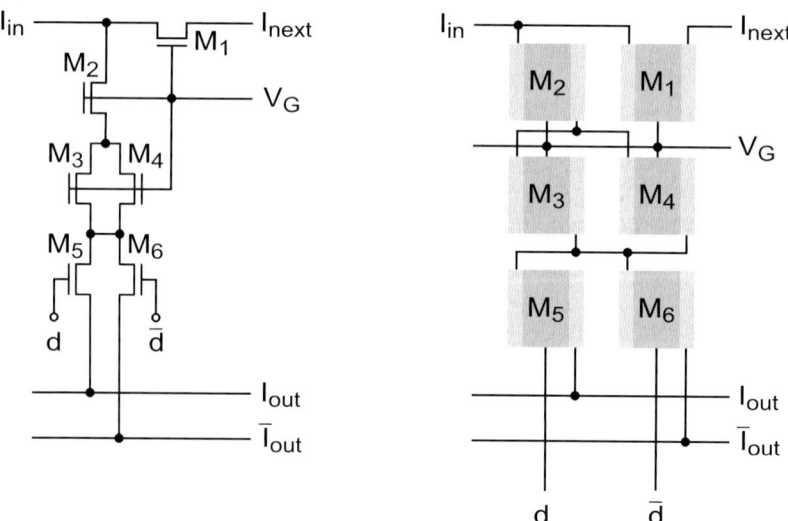

Figure 50. Layout overview of one stage of a M/2.5M converter

The input current I_{in} propagates to next stage I_{next} on top of the structure, whereas the common gate voltage line is in the middle and the output current

Chapter 3: Digital compensation circuits 73

collectors are at the bottom. The digital control inputs are extracted at the bottom.

This layout has the advantage of allowing the cells to be adjoined regularly to build up the ladder. Replicas of this 2 by 3 matrix can be abutted horizontally, leading to a simple and compact layout.

The M/3M ladder can be realized similarly, by simply transforming transistor M_4 into a dummy.

11.7 Measurements

This section presents measurement results of a test-chip integrated in a 0.8µm process and containing 6 different $M/2^+M$ ladders with 16 stages: 4 differently sized M/2.5M and 2 different M/3M ladders. Figure 51 shows a micrograph of this circuit and table 8 presents the dimensions W and L of the unity transistor used in each network.

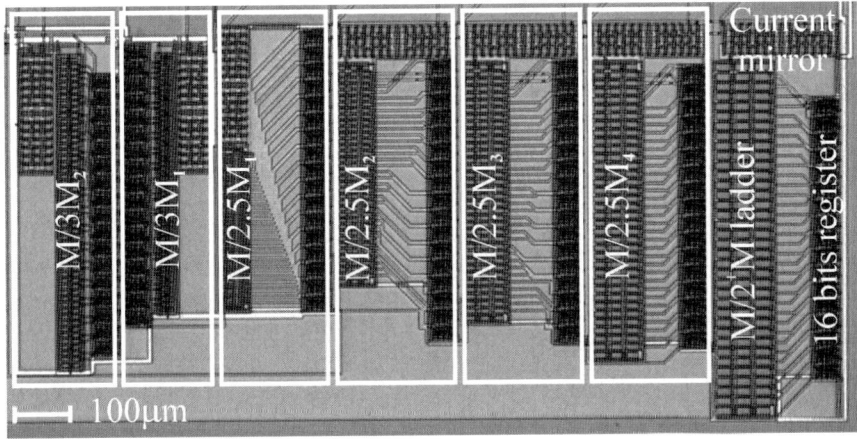

Figure 51. $M/2^+M$ test-chip micrograph

The transistor sizes used in this test-chip are voluntarily large, in order to study the influence of the circuit area on the dispersion of the current division. The result of this analysis is that even the smallest devices can be used without violating the working condition of the successive approximations algorithm.

The dispersion is measured on 10 samples, which is too few to perform a reliable statistical analysis. The results presented below only allow a qualitative interpretation.

Table 8. M/2⁺M test-chip ladder characteristics

Ladder name	W	L
M/2.5M$_1$	6	3
M/2.5M$_2$	12	6
M/2.5M$_3$	16	8
M/2.5M$_4$	20	10
M/3M$_1$	8	4
M/3M$_2$	12	6

The first result is that none of the converters in any sample violates the working condition. At every stage, the sum of the currents of the least significant bits is always higher than the current of the corresponding bit.

Table 9 presents the mean current division factor ρ measured for each ladder, compared with the theoretical value ρ_{stat}.

Table 9. M/2⁺M current division measurement

Ladder name	ρ	ρ_{stat}
M/2.5M$_1$	0.88	0.86
M/2.5M$_2$	0.86	
M/2.5M$_3$	0.87	
M/2.5M$_4$	0.87	
M/3M$_1$	0.77	0.77
M/3M$_2$	0.78	

These measurements are close to the theoretically expected results and confirm that the ladders divide the currents adequately. Figure 52 shows the mean standard deviation of the current division factor ρ for the M/2.5M ladders as a function of the device area. The x-axis corresponds to the inverse of the square root of the area, in order to allow a visual interpretation of:

$$\sigma(\rho) = \frac{A_\rho}{\sqrt{W \cdot L}} \qquad (3.84)$$

Chapter 3: Digital compensation circuits

which is derived from equation 2.1.

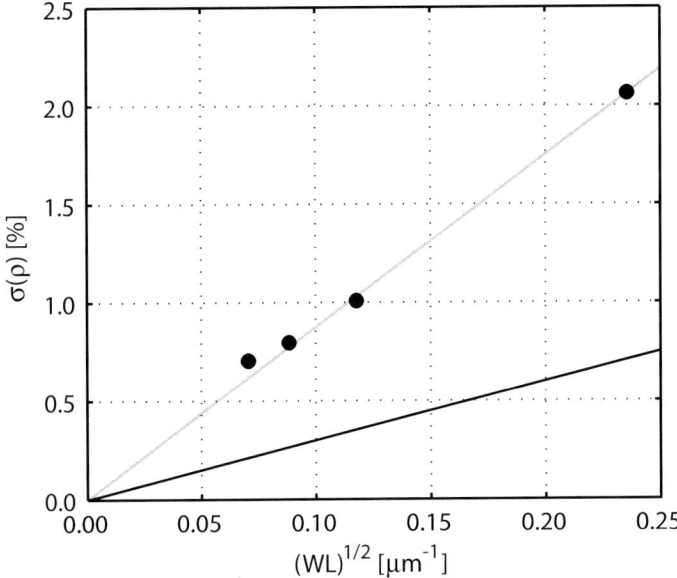

Figure 52. Standard deviation of the current division in M/2.5M ladders

In figure 52, the slope corresponds directly to the matching parameter A_p, which is the standard deviation for unity-area devices. The expected matching is the black line, whereas the measured deviations are the black points. The grey line is the fitting of the measurements for small areas (large values on the x-axis).

The first observation is that the measured deviation is about 3 times larger than the prediction from the technology matching parameters. This is explained by the fact that the elements in the ladder are not really matched. The layout is based on a regular grid, but several matching rules are violated. For instance, the layout is not common-centroid. If there is a gradient during processing, the devices are not affected evenly.

The second observation is that for large areas, the deviation becomes even larger than 3 times the prediction. This is because the distance between the transistors also increases when the devices are larger. In this case, the current division relies no more on mismatch, but on the relative tolerance of the components, which is not a function of surface and that is clearly non-zero.

The dispersion of the current division factor ρ also depends on the considered stage in the ladder. Figure 53 shows the standard deviation as a function of the stage number for the largest implemented M/2.5M ladder (M/2.5M_4).

The black points correspond to the measurements, and the grey curve is a fitting of these results.

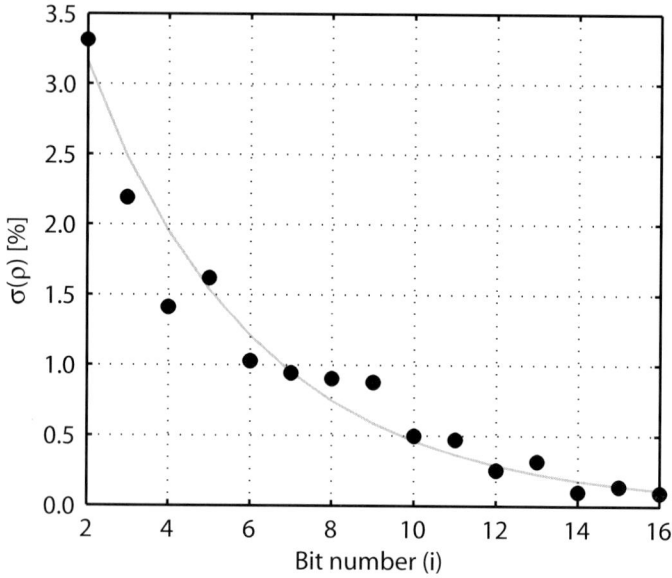

Figure 53. Standard deviation of ρ in each stage of the M/2.5M$_4$ ladder

In the last stages of the ladder, the current division depends on the relative characteristics of only a few components, whereas the equivalent resistance of the ladder in the first stages is built up from all the devices of lower stages. For this reason, the dispersion is higher for the first bits.

In this case, one can note that in stage number 2, the standard deviation is close to the acceptable limit in careful designs. In fact, considering a deviation of $3\sigma \cong 10\%$ from the nominal value $\rho \cong 0.87$, the worst-case is close to 1, which is the limit for the working condition (equation 3.48).

In the case of M/3M converters, this is less problematic. Figure 54 shows the same standard deviation as a function of the stage number, but this time for the smallest implemented M/3M ladder (M/3M$_1$). In this case, the nominal value $\rho \cong 0.77$ increased by $3\sigma \cong 18\%$ is still far from the 1 limit.

Chapter 3: Digital compensation circuits

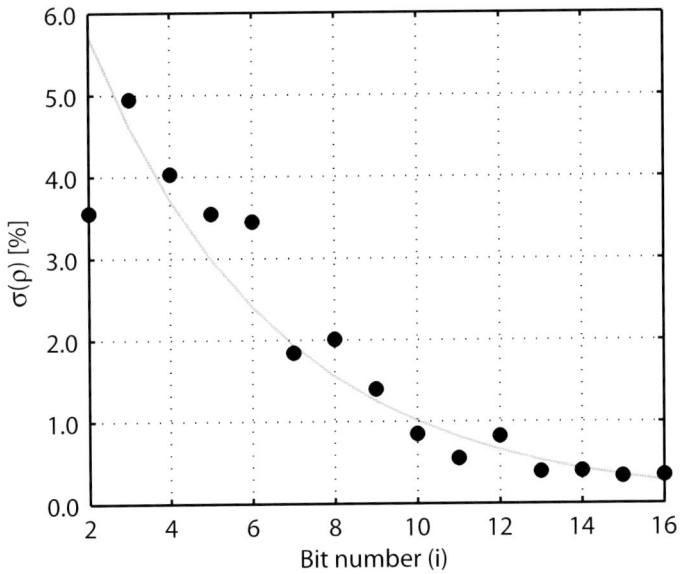

Figure 54. Standard deviation of ρ in each stage of the M/3M$_1$ ladder

In the case of a M/3M ladder, a very small unity device can thus be implemented, even smaller than the W/L = 8/4 used here. If, on the contrary, a M/2.5M ladder is preferred, the transistor sizes should be somewhat larger to ensure proper current division within the 3σ interval.

The important learning presented here is that very small devices (almost minimal) can be used to design sub-binary M/2$^+$M converters. The 16 stages M/3M$_1$ converter for instance, even if it is still larger than necessary, occupies only 0.03 mm^2. This is comparable to the area of the current mirror implementation [23] presented in section 6.1.

12 COMPARISON

This chapter presented different digital-to-analog converter structures that can be used with successive approximations algorithms. The most important conclusion is that sub-binary converters are especially well suited for this application, because they are simple to design and use little circuit area. Among the different sub-binary converter structures, the current-mode M/2$^+$M and current-mirror DACs are the most generic. Their use in a complete design methodology is presented in the next chapter. Concerning M/2$^+$M ladders, the simplicity and regularity of their structure even allows fully-automatic layout. Except for the above-mentioned considerations, all the sub-binary converters presented in this chapter have similar performances and design guidelines.

13 LINEAR DACS BASED ON M/2$^+$M CONVERTERS

The fact that sub-binary have a radix lower than 2 is advantageous for successive approximations. It allows the design of very low-area converters with low-precision devices. Unfortunately, they are not linear and cannot be used directly as conventional DACs. However, using the digital self-calibration technique presented in this section [37], their characteristics can be rectified, allowing the design of high-precision DACs.

The presented technique does not require precise additional analog components, and only little digital correction circuit area. Conserving the same philosophy as M/2$^+$M converters it thus permits to realize high-performance DACs, even with manufacturing processes where the quality of analog devices is poor.

13.1 Principle

The input/output characteristics of sub-binary digital-to-analog converters are non-monotonic. To ensure that there is no missing code, redundant codes are introduced. This property is not a problem for the successive approximations algorithm, but is not tolerable if linear characteristics are expected. However, by using two algorithms presented in sections 13.2 and 13.3, the redundant codes can be removed and the characteristics of the DAC made linear and monotonic.

Figure 55 shows the input/output characteristics of a 4-bits uncalibrated sub-binary converter.

Chapter 3: Digital compensation circuits

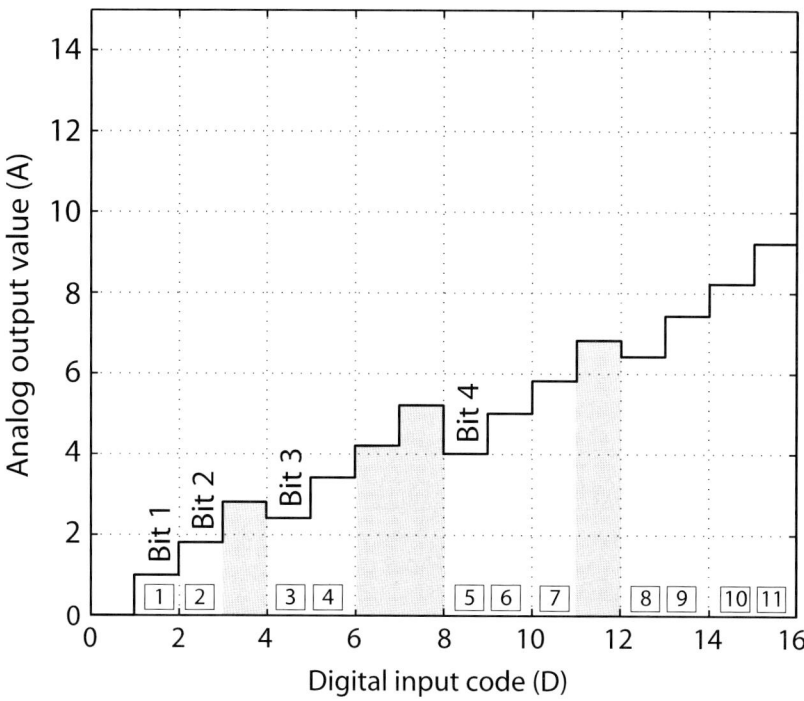

Figure 55. Input/output characteristics before calibration

If the shaded codes are removed and only the framed ones are kept, the characteristics become monotonic, as shown in figure 56. The removed codes are those which generate higher output values than higher input codes, and thus create non-monotonicities. It is noteworthy that the identification of these codes can be done using the DAC itself, and that no additional element is necessary except a current comparator and an appropriate algorithm to identify the problematic codes.

The calibrated input/output characteristics is not only monotonic, but also remarkably linear. Although the step sizes are random, the structure of their repetition constructs a surprisingly regular slope. The principle is similar to the *accuracy bootstrapping* method described in [38], [39] and [40], which is used to self-calibrate pipelined ADCs.

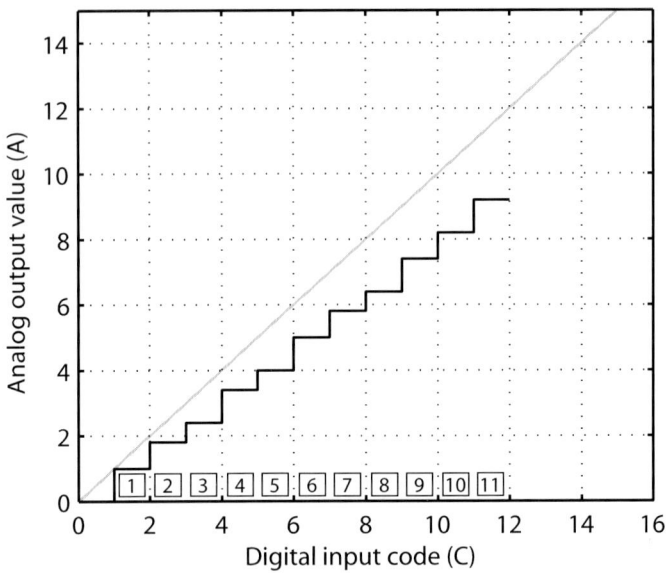

Figure 56. Input/output characteristics after calibration

Another property of the calibrated input/output characteristics is that its slope is inferior to the ideal unity function plotted in grey in figure 56. This is because the step amplitude is randomly comprised in the [0, 1] interval. This is not a problem, however, since in current-mode $M/2^+M$ DACs the bias current has a multiplicative effect on all the output codes. By adjusting the bias current, it is thus possible to correct the slope.

To obtain the calibrated characteristics, the circuit of figure 57 is implemented. The system consists in a main DAC which is calibrated digitally to obtain the input/output characteristics of figure 56, and whose slope is adjusted to a unity value. The main purpose of the digital calibration is the *identification* and *removal* of the codes that cause non-monotonicities. These two tasks are performed by two algorithms, which interact through a calibration table.

The calibration algorithm presented in section 13.2 is used before circuit operation, at power-up for instance, to identify the redundant codes. It stores their location in the calibration table. It also adjusts the full scale of the main DAC to be equal to the external reference by the means of the full scale adjust DAC. Once calibrated, the DAC is ready for normal use.

Chapter 3: Digital compensation circuits

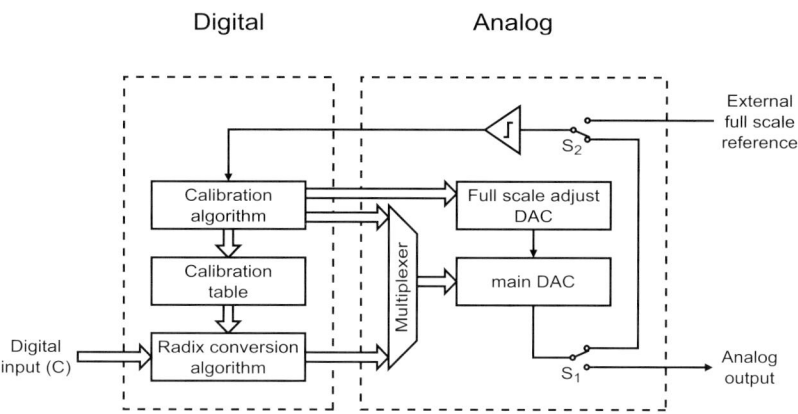

Figure 57. DAC system architecture

For each digital input code, the second algorithm, the radix conversion algorithm presented in section 13.3, computes the appropriate digital code for the main DAC, using the informations contained in the calibration table.

13.2 Calibration algorithm

The digital calibration is in charge of filling in the calibration table before circuit operation. Its goal is to identify the codes that cause the DAC to be non-monotonic (the shaded codes in figure 55). To configure the circuit for calibration, the switches S_1 and S_2 are positioned as shown in figure 57.

It is noteworthy that the codes generating non-monotonicities are always located just before the transition of a bit from 0 to 1. In figure 55, the transition of the third bit (code 4) is preceded by one higher value corresponding to a code to be removed. The same problem is repeated before code 12, where bit 3 also toggles from 0 to 1. The transition of the MSB (bit 4, code 8) is even worse since two redundant codes are located just before.

The calibration algorithm must identify the redundant codes that precede the transition of a bit i from 0 to 1. More precisely, it has to find the last (highest) code C_{limit} before the transition that is still inferior to the value b_i of the corresponding bit, as shown in figure 58.

C_{limit} is identified by a successive approximations algorithm which compares the lower codes to b_i. The algorithm indeed finds the highest code that is inferior (by no more than 1 LSB), as stated by equations 3.3 and 3.4.

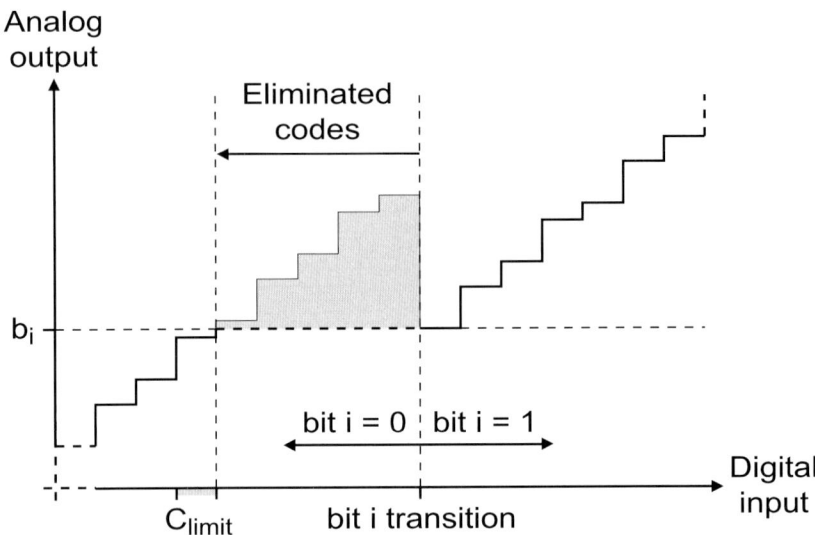

Figure 58. DAC calibration principle

Using this procedure, every bit is calibrated successively, starting from the LSB (i = 1) and ending with the MSB (i = n). Each bit is estimated using the already calibrated least significant bits and the analog comparator.

For each bit i, the algorithm stores in the calibration table its estimated weight w_i. Arbitrarily, w_1 is set to 1. The further weights are calculated for each bit as the sum of the already calculated weights of the bits that are kept by the successive approximations algorithm, plus one. The values w_i can thus be viewed as the estimated weights of the bits b_i in function of the other least significant bits. But the w_i are also equal to the total number of monotonic steps until the transition of a bit i from 0 to 1 (including the transition step itself).

The first step of the calibration algorithm (figure 59) is to calibrate arbitrarily the LSB to be 1 elementary step. Then, the outer loop calibrates each bit successively by using the successive approximations algorithm (inner loop). The comparisons allowing the algorithm to decide which bits have to be kept are performed by the single-input analog current comparator.

Chapter 3: Digital compensation circuits

```
w₁ = 1
for i = 2 to n
    wᵢ = 1
    b = 0
    for j = i - 1 downto 1
        if b + bⱼ < bᵢ
            b = b + bⱼ
            wᵢ = wᵢ + wⱼ
        end if
    end for
end for
```

Figure 59. DAC calibration algorithm

At the end of the successive approximations (inner loop), the value b corresponds to the output of the DAC when its digital input code is C_{limit}:

$$b = \sum_{j=1}^{i-1} d_j b_j = b_i - \varepsilon \qquad (3.85)$$

where d_j is the digital input bit i of the DAC, and $\varepsilon \in [0, 1]$ the difference between the two codes, and thus step amplitude after calibration. Equation 3.85 is the consequence of equations 3.3 and 3.4.

The final calculated weight w_i is:

$$w_i = 1 + \sum_{j=1}^{i-1} d_j w_j \qquad (3.86)$$

Table 10 shows the result of the calibration algorithm for the example of figure 55. It can be verified in figure 55 that the weights w_i are the number of monotonic steps (non-shaded codes) until the corresponding bit transition (including itself), since they are equal to the framed numbers which count them. This information is thus indeed the estimated weight of b_i in function of the least significant bits, as shown in figure 56. As expected, the estimated weights w_i are a sub-binary radix number system, exactly as the b_i themselves. It is also noteworthy that if an ideal binary DAC is calibrated with this algorithm, the weights are estimated perfectly ($w_i = b_i = 2^{i-1} \ \forall i$).

Table 10. Calibration table for the example of figure 55

bit number (i)	calibrated weight (w_i)
1	1
2	2
3	3
4	5

The complexity of the calibration algorithm is in the order of $O(n^2)$, because there are two loops enclosed one in another, both on the number of bits n of the DAC.

Once the calibration table is calculated, the full scale of the main DAC is adjusted by comparing it to the external reference through switch S_2. The adjustment is done by a successive approximations algorithm controlling the full scale DAC. Once the calibration is finished, switch S_1 is toggled and the DAC is ready for normal use.

13.3 Radix conversion algorithm

To allow the elimination of the redundant codes, each digital input code C of the system must be processed by the radix conversion algorithm to find the corresponding digital control code for the main DAC. The performed calculation is indeed a radix conversion, since the calibration table contains the (estimated) weights of the bits of the main DAC and that C is the desired monotonic step to be generated.

The radix conversion algorithm (figure 60) performs successive approximations in the digital domain, basing the decisions on the values in the calibration table. The variable e is the weight that still has to be estimated. At the end of the algorithm, e = 0 and the output of the converter is the Cth monotonic step, which is the required value. It can be verified that the digital input sequence from 0 to 11 produces the characteristics of figure 56 if the codes are processed by the radix conversion algorithm.

Chapter 3: Digital compensation circuits 85

```
e = C
for i = n downto 1
    if e ≥ wᵢ then
        e = e - wᵢ
        dᵢ = 1
    else
        dᵢ = 0
    end if
end for
```

Figure 60. DAC radix conversion algorithm

In terms of complexity, the algorithm is fast, since it performs only n (the number of bits of the DAC) comparisons and subtractions. This signifies that the digital circuit must operate n times faster than the analog part, because it must convert each input code C into the digital control word D of the DAC in real time.

13.4 Digital circuit implementation

Both calibration and compensation algorithms work in a very similar manner (by successive approximations), use the same datas (the calibration table) and perform the same operations (additions and subtractions). Although the exact sequence of operations differs, the core of both algorithms is similar. This implies that the circuit implementing the data processing unit and the successive approximations control logic can be shared. Figure 61 shows the necessary circuits.

Each algorithm is implemented by a small dedicated control logic circuit which performs the corresponding sequence of operations using the shared processing blocks and the implementation of the successive approximations algorithm. In calibration mode, the calibration algorithm takes control of the processing unit, whereas it is operated by the radix conversion system during normal operation.

The calibration table is an array of n registers, one for each bit of the DAC. Because the calibrated weights are a sub-binary number system, the matrix necessarily contains only 0s in the upper diagonal part. The number of necessary memory cells in the calibration table is thus divided by two.

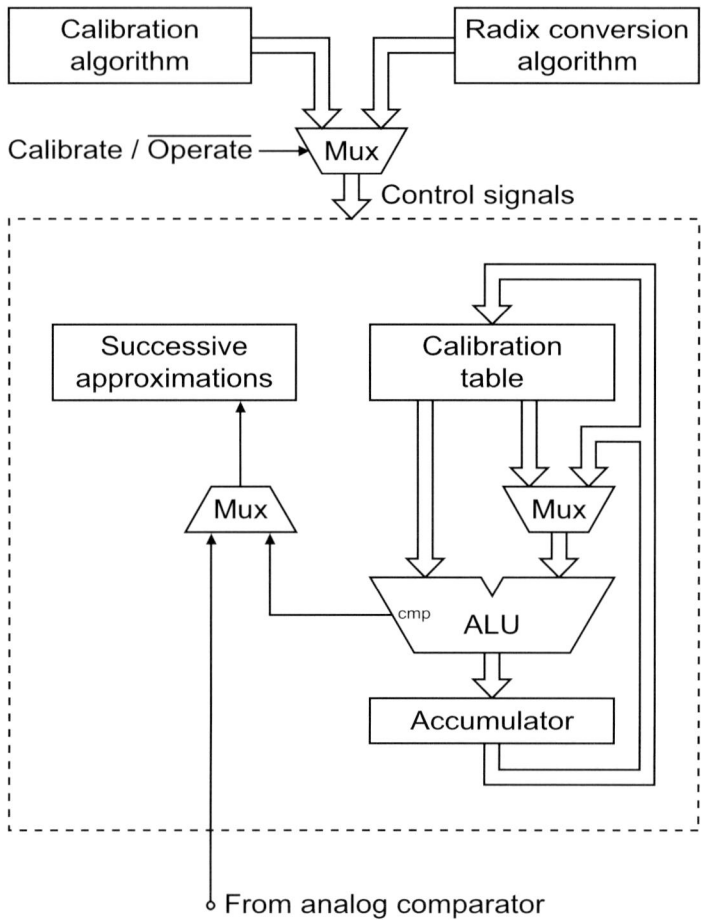

Figure 61. Digital circuit implementation

The processing unit is composed of an ALU and an accumulator. It must only perform additions, subtractions and comparisons and can thus be implemented using only a n-bits inverter (for 2's complement subtraction) and an n-bits full adder. The inputs of the ALU are two lines (two different w_i) of the calibration table for the calibration algorithm, whilst one w_i and the accumulator are the operands during radix conversion. This implies the use of a multiplexer for the second input of the ALU, but also two read channels and one write port for the calibration table.

The successive approximations algorithm is operated differently by both algorithms. In fact, the decision input is the output of the analog comparator during calibration, whereas it is the digital comparison signal cmp coming from the ALU during radix conversion. The second difference is the start bit of the algorithm since, during calibration, it is not the MSB but the most significant bit after the currently calibrated bit.

The complexity of this digital control logic is low. The complete digital circuit has been designed in VHDL and a layout generated in a 0.35 μm technology using an automatic synthesis tool. The total area of the circuit is 0.3 mm^2.

13.5 Analog circuit implementation

The main DAC and the full scale DAC are M/2$^+$M converters, either M/2.5M or M/3M. Their implementation is discussed in section 11.

One special consideration in the case of a linear DAC is the current collector, which cannot be a simple current mirror (section 11.4) as for successive approximations. In fact, the current division in each branch of the converter depends on the voltage in the two current collector nodes. If a linear DAC and thus a linear current summation is desired, it is mandatory that these voltages do not depend on the collected current and remain constant and equal.

A simple solution to stabilize the voltage is to use a transresistance amplifier as presented in figure 62.

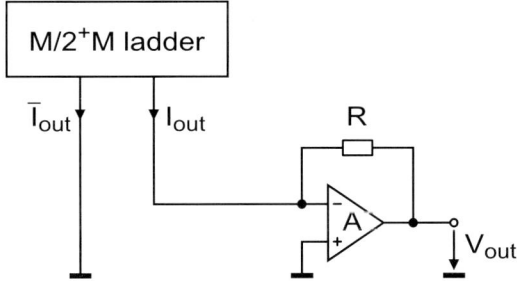

Figure 62. Transresistance current collector

If a differential output is desired, a second transresistance amplifier is connected to the complementary current output instead of dumping it directly into the ground.

To guarantee the stability of the voltage in the current collector node N_C, the open-loop gain A of the amplifier is made high. The current is converted into a voltage by the resistor R, and the output voltage is:

$$V_{out} = -RI_{out} \quad (3.87)$$

The offset of the amplifier also affects the performance since the feedback copies it (with an opposite sign) to N_C. In high-resolution DACs, the offset of the amplifier can be calibrated (by autozero for example) in order to meet the requirements in terms of maximum voltage difference between the current collector nodes.

The transresistance circuit converts the current into a voltage. If a current output is desired, or to overcome the limitations of this topology in high-precision designs [29], the regulated cascode of figure 63 can be used.

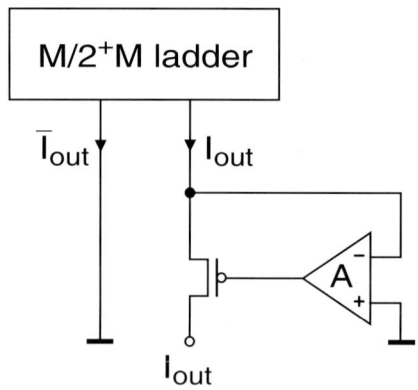

Figure 63. Regulated cascode current collector

The disadvantage of this structure is the difficulty to guarantee the stability of the feedback loop of the amplifier for the complete range of output currents I_{out}, from the LSB to the full scale. This limitation can be overcome by permanently adding an offset bias current to I_{out}, in order to guarantee a minimum drain current for the cascode transistor. A more elaborate version of current collector can be found in [29].

The current comparator used by the calibration algorithm is presented in figure 64.

Chapter 3: Digital compensation circuits

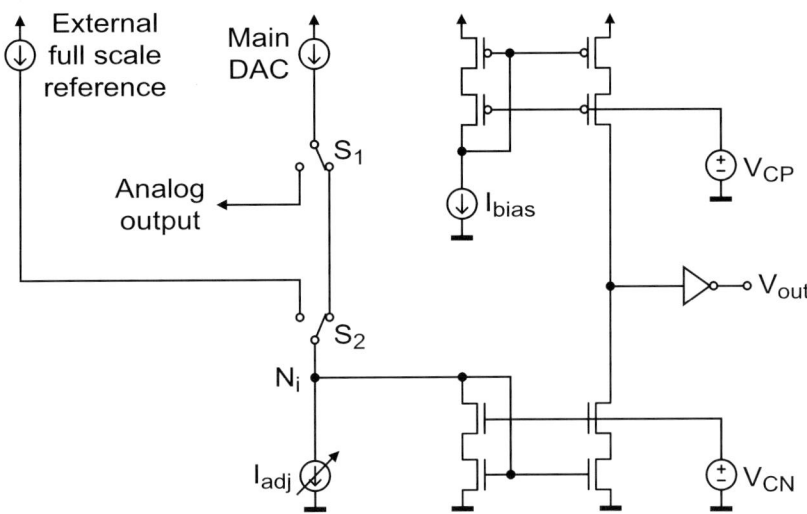

Figure 64. Single-input current comparator

Because the two currents to be compared are both generated by the main DAC during calibration, the comparator has a single input N_i where the values to be compared are presented successively. The sensitivity of the comparator must allow to detect a current difference as low as 1 LSB, in the range between 0 and the full scale current. This is achieved by adding an additional current source I_{adj} (another sub-binary converter), which is adjusted by successive approximations until its value is almost equal to the current generated by the main DAC. The small residual current flows through the high-gain cascode stage and brings the output V_{out} close to its toggling point. The second current is then applied instead of the first one and the comparison result is determined from the presence or absence of transition of V_{out}.

Using this technique, it is possible to perform very precise current comparisons in a large range. Furthermore, since the cascode current mirror is used only to compare currents, its linearity and precision are not important. It can thus be realized on a small circuit area and using low-precision components, like the other analog blocks in the circuit.

The complete analog circuit using a differential output with regulated cascode current collectors has been implemented in a 0.8 µm technology. The micrograph is presented in figure 65.

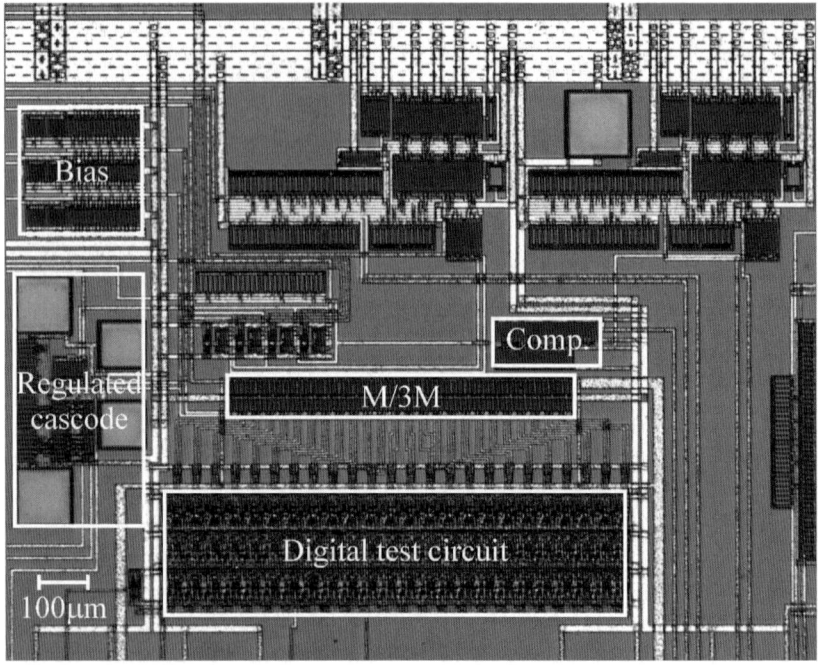

Figure 65. DAC micrograph

13.6 Compensation of temperature variations

The calibration of the DAC with the calibration algorithm of section 13.2 can be performed one single time at power-up. However, the fact that the components of the DAC are small and poorly matched causes the current division to be sensitive to temperature variations. In high-resolution converters, this limits the performance if the circuit is not re-calibrated at regular time intervals.

In applications where the output is not continuous and at low frequencies, the DAC may be re-calibrated between the generation of two consecutive output samples. Otherwise, the analog circuit can be duplicated and ping-pong can be performed: During the calibration of one DAC, the other is used for conversion and the roles are interchanged at the end of the adjustment. This duplication is not an issue because the analog circuit area is very small.

13.7 Comparison with other self-calibrated converters

The self-calibration methodology proposed in this section [37] allows the realization of high-precision DACs using low-precision analog components. By digitally calibrating the sub-binary DAC, the duplicate codes are removed and the converter can be used as a conventional DAC. The calibration is performed using the sub-binary DAC itself (self-calibration), and does not require additional circuits.

The presented architecture is attractive because the analog part is simple to design and robust to imperfections. It does not need precise analog components, like a calibration ADC as in [41] and [21]. A single current comparator with relaxed design constraints is sufficient. Furthermore, the compensation is purely digital and there is no analog calibration as in [42] and [43].

The reduced analog circuit constraints allow easy design and retargeting, even with fully digital technologies. The automatic conception of both analog and digital parts is also possible. All these advantages, along with the very little total circuit area, make this new topology promising for the realization of high-performance digital-to-analog converters.

14 CONCLUSION

Sub-binary converters are a must for successive approximations. They significantly relax the design constraints by achieving an arbitrarily high precision even with poor components. Furthermore, they can be implemented with very little circuit area. The current-mode sub-binary converters are particularly attractive, because many circuit imperfections can be compensated by injecting a compensation current in a well-chosen node. A complete methodology using sub-binary current-mode converters is presented in the next chapter. Finally, using appropriate calibration algorithms, it is even possible to use sub-binary converters as conventional DACs.

Chapter 4

Methodology for current-mode digital compensation of analog circuits

Chapter 3 presents the successive approximations algorithm, as well as the $M/2^+M$ sub-binary converters which are recommended to be used in conjunction. In this chapter, a complete methodology using these current-mode digital correction circuits for improving the performance of analog systems is presented. The explanation is based on the example of the offset compensation of an operational amplifier. Another example, the calibration of a SOI 1T DRAM current reference, is presented at the end of the chapter.

1 INTRODUCTION

The successive approximations algorithm, combined with a current-mode digital-to-analog converter like a $M/2^+M$ for instance, allows to compensate a large variety of analog circuit imperfections. This chapter presents a complete methodology based on such digital correction circuits which inject a compensation current. It ranges from the choice of the most adapted compensation to the verification of its efficiency using adapted simulation tools. The example of the offset correction of an operational amplifier is used throughout the chapter for simplicity and coherence reasons. The application of the methodology to another circuit, the SOI 1T DRAM current reference, is also presented.

2 TWO-STAGE MILLER OPERATIONAL AMPLIFIER

This section presents the operational amplifier example, which is used in this chapter for demonstration purposes. The two-stage Miller topology is chosen for its simplicity. Because this structure is well-known, the focus is set on the compensation methodology rather than on the amplifier characteristics and design constraints. But this topology is used for demonstration reasons only. In practice, other more efficient structures are chosen to overcome the limitations of the two-stage Miller operational amplifier. These are principally a limited gain, a poor power-supply rejection ratio and the difficulty to control its stability [7].

Other structures like the cascode operational amplifier are usually considered to improve circuit performance. The design complexity of these more elaborate topologies can be reduced by using appropriate CAD tools, like the Procedural Analog Design (PAD) tool [44]. By splitting the design procedure into consistent sub-block sizing, PAD enables the designer to gain insight at the same time into transistor-level and circuit-level parameters. It also allows to understand their interdependence and the design trade-offs.

Figure 66 shows the schematic of the well-known two-stage Miller operational amplifier [7]. C_L is the output load capacitor, and C_C the compensation capacitor that is necessary to stabilize the circuit by placing the dominant pole to increase the phase margin to a sufficient value. The series resistor R_Z is optional, but allows a better control over the zero introduced by the compensation capacitor. Usually, this resistor is realized using a MOS transistor.

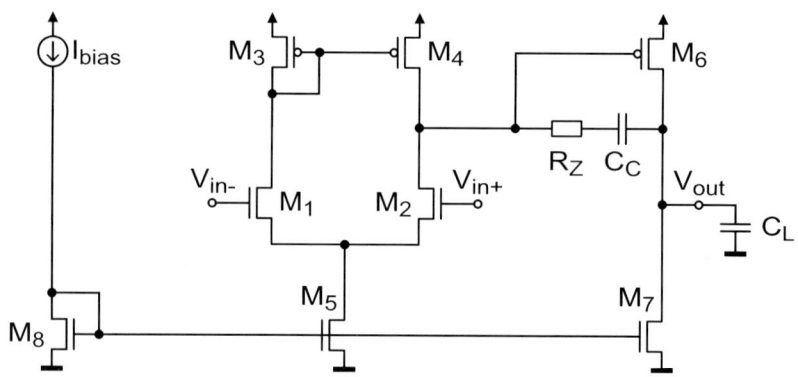

Figure 66. Two-stage Miller operational amplifier

Figure 67 shows the small-signal model of this circuit.

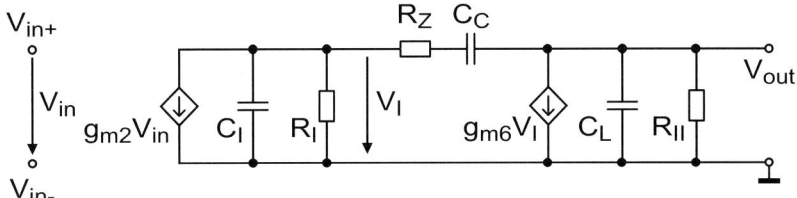

Figure 67. Small-signal model of the two-stage amplifier

C_I is the parasitic capacitance seen from the drain of M_2 and M_4 to the ground. R_I is the impedance in this node and is equal to:

$$R_I = \frac{1}{g_{ds2} + g_{ds4}} \quad (4.1)$$

Where g_{ds} is the channel conductance, which is calculated as:

$$g_{ds} = \lambda I_D \quad (4.2)$$

λ is the channel-length modulation factor and I_D the bias drain current. λ depends on the length of the transistor and can be approximated by:

$$\lambda \cong \frac{\alpha}{L} \quad (4.3)$$

where α is a technology-dependent parameter.

The transconductance g_m of a saturated transistor (in strong inversion) is equal to:

$$g_m = \sqrt{2\mu C_{ox} \frac{W}{L} I_D} \quad (4.4)$$

where μ is the electron/hole mobility and C_{ox} the oxide capacitance.

In the second stage, the parasitic capacitance is usually much smaller than the output load capacitance C_L and is neglected for this reason. The impedance at the output is calculated as:

$$R_{II} = \frac{1}{g_{ds6} + g_{ds7}} \quad (4.5)$$

This equation does not account for an output load. If the output is loaded by a resistive charge R_L, R_{II} becomes:

$$R_{II} = \frac{1}{g_{ds6} + g_{ds7} + (1/R_L)} \tag{4.6}$$

Table 11 summarizes the principal characteristics of the two-stage Miller operational amplifier.

Table 11. Characteristics of the two-stage Miller operational amplifier

Parameter	Value
Unity-gain bandwidth	$BW = \dfrac{g_{m2}}{C_C}$
Dominant pole	$p_1 = \dfrac{(g_{ds2} + g_{ds4})(g_{ds6} + g_{ds7})}{g_{m6} C_C}$
Second pole	$p_2 = \dfrac{g_{m6}}{C_L}$
Right half-plane zero	$z = \dfrac{g_{m6}}{C_C}$ $(R_Z = 0)$ $z = \dfrac{1}{C_C(1/g_{m6} - R_Z)}$ $(R_Z \neq 0)$
First stage gain	$A_I = \dfrac{g_{m2}}{g_{ds2} + g_{ds4}}$
Second stage gain	$A_{II} = \dfrac{g_{m6}}{g_{ds6} + g_{ds7}}$
Total gain	$A = A_I A_{II}$

3 COMPENSATION CURRENT TECHNIQUE

To allow the digital compensation of an imperfection in an analog circuit by the injection of a compensation current, two nodes must be identified in the analog system: a *detection node* and a *compensation node*. Furthermore, a *detection configuration* of the compensated analog circuit must be found.

Chapter 4: Digital compensation of analog circuits 97

The observation of the detection node allows to determine whether the imperfection to be corrected is lower or higher than the error-free nominal value to be reached by calibration. For example, the offset of an operational amplifier is positive or negative around the nominal value 0. Based on the information gathered in the detection node, an adequate decision is taken to compensate the imperfection by increasing it if it is lower than expected or decreasing it if it is higher than the nominal value.

This correction is done by injecting a compensation current in an appropriate compensation node of the analog circuit. The compensation node is chosen for its property to convert the injected compensation current into a reduction/increase of the imperfection. In the ideal case, no other parameter should be affected by the compensation current. In practice, the correction node is chosen for the high correlation between the injected current and the imperfection, and the low correlation between the compensation current and any other parameter.

Both the observation node and the compensation node can also be differential, i.e. the detection and/or correction can also be a function of the difference between two signals rather than a function of a single absolute value.

The detection configuration is the condition in which the circuit imperfection can be observed in the detection node. In some cases, this is possible during normal circuit operation, without modifying the circuit topology and interrupting signal processing. But in many cases, a different circuit configuration is necessary to measure the imperfection in the detection node.

3.1 Detection configuration

In this section, two different versions of the same amplifier circuit are presented. The fist one allows continuous-time compensation of the offset, whereas the second one requires the interruption of normal circuit operation to perform the compensation. Both topologies are briefly introduced below. In the next sections, they are analyzed in detail and their advantages and drawbacks are discussed.

Suppose that the two-stage Miller operational amplifier of section 2 is used in the configuration of figure 68. The shaded triangle represents the circuit of figure 66. The voltage source V_O models the input offset voltage of the amplifier, whose open-loop gain is A.

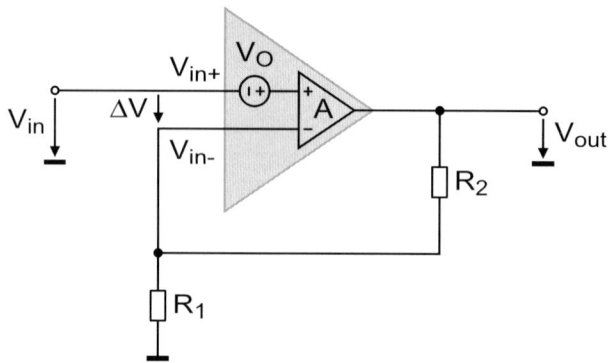

Figure 68. Offset detection in the closed-loop configuration

The output voltage V_{out} is calculated as:

$$V_{out} = A(V_{in+} - V_{in-} + V_O) = A(V_{in} - V_{in-} + V_O) \qquad (4.7)$$

The resistive ladder performs feedback:

$$V_{in-} = V_{out}\frac{R_1}{R_1 + R_2} \qquad (4.8)$$

Replacing equation 4.8 in 4.7 gives:

$$V_{out} = \frac{1}{\frac{R_1}{R_1 + R_2} + \frac{1}{A}}(V_{in} + V_O) = A_{closed\text{-}loop}(V_{in} + V_O) \qquad (4.9)$$

The input voltage and the input offset voltage are amplified by the same constant, the closed-loop gain $A_{closed\text{-}loop}$ which is in the ideal case ($A = \infty$) equal to:

$$A_{closed\text{-}loop} = \frac{1}{\frac{R_1}{R_1 + R_2} + \frac{1}{A}} \cong \frac{R_1 + R_2}{R_1} \qquad (4.10)$$

In equation 4.7, it is also possible to identify ΔV:

Chapter 4: Digital compensation of analog circuits

$$\Delta V = V_{in+} - V_{in-} = \frac{V_{out}}{A} - V_O \qquad (4.11)$$

Replacing 4.9 in 4.11 gives:

$$\begin{aligned}\Delta V &= \frac{A_{closed-loop}}{A}(V_{in} + V_O) - V_O \\ &= \frac{A_{closed-loop}}{A}V_{in} + \left(\frac{A_{closed-loop}}{A} - 1\right)V_O\end{aligned} \qquad (4.12)$$

Usually, the closed-loop gain $A_{closed-loop}$ is much smaller than the open-loop gain A. In this case, equation 4.12 simplifies to:

$$\Delta V \cong -V_O \qquad (4.13)$$

The offset voltage can be observed directly between the input terminals V_{in+} and V_{in-} of the operational amplifier. A second-order component, due to the fact that the open-loop gain of the operational amplifier is not infinite, adds to this value a fraction of the input signal and an additional fraction of the offset voltage.

It is possible to take advantage of the property of equation 4.13 to compensate the offset of the operational amplifier continuously, without interrupting normal circuit operation. This can be done by sensing the voltage difference ΔV between the two inputs of the operational amplifier to be compensated. An analog circuit performing this task is described in [45], [6] and [46]. The compensation can also be performed digitally, by connecting a comparator in parallel with the amplifier to be compensated. The disadvantage, however, is the low level of the detection signal, as detailed in section 3.2. Another shortcoming of this circuit is the additional circuitry necessary to operate the compensation in continuous time (section 3.6), because the direct use of a successive approximations algorithm creates large temporary offsets.

Another detection configuration for the same circuit is presented in figure 69. The switch S_{mode} allows to configure the circuit in two different modes. In the down position, the amplifier is in normal operation mode and corresponds exactly to the circuit of figure 68. In the up position (as in the figure), the circuit is in its detection configuration, where the offset is detected and compensated.

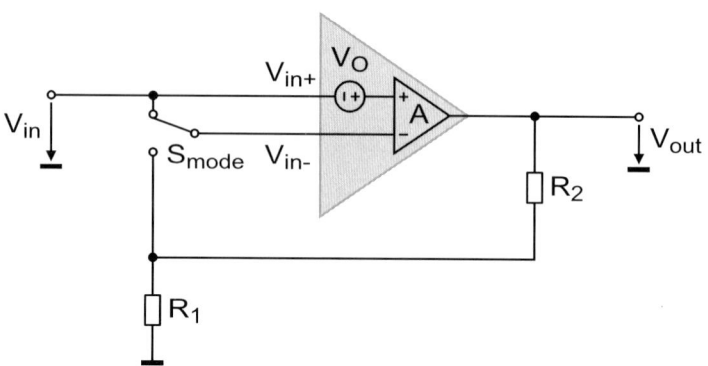

Figure 69. Offset detection in the open-loop configuration

In compensation mode, both inputs of the operational amplifier are connected together to the input signal. The amplifier operates as a comparator and its output is:

$$V_{out} = A(V_{in+} - V_{in-} + V_O) = AV_O \qquad (4.14)$$

The offset voltage is measurable directly at the output, multiplied by the open-loop gain of the amplifier. This topology is better suited for the detection of the offset because it presents a much higher detection signal level. On the other hand, it requires the interruption of normal operation. This problem can be solved by using the ping-pong technique presented in chapter 2, section 6.

3.2 Detection node

A detailed analysis of the node where the imperfection is sensed is presented below for both the closed-loop and open-loop configurations of the last section.

In the closed-loop configuration, as indicated by equation 4.13, it is possible to sense the offset of the operational amplifier by connecting a comparator to its inputs. The corresponding circuit is shown in figure 70. In this circuit, the sensing is differential between the voltage in the two detection nodes V_{in+} and V_{in-}. The *detection signal* δ is derived from equation 4.13:

$$\delta = V_{in+} - V_{in-} \cong -V_O \qquad (4.15)$$

Chapter 4: Digital compensation of analog circuits

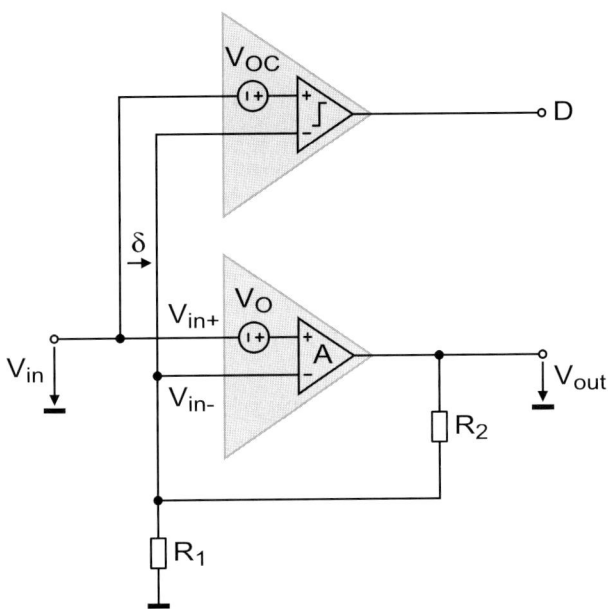

Figure 70. Offset measurement in the closed-loop configuration

Let's suppose that the comparator is ideal and that its offset voltage V_{OC} is null. In this case, its output D controls the decision to be taken by the digital correction circuit to compensate the offset of the amplifier. In this text, the following convention is used: D = 1 if the voltage difference between the positive and negative inputs of the comparator is higher than 0, and D = 0 otherwise. The digital circuitry must act on the compensation node in order to increase the offset if D = 1 ($\delta > 0$; $V_O < 0$) and decrease it if D = 0 ($\delta \leq 0$; $V_O \geq 0$). If this is done with an infinite resolution, $\delta = 0$ and the final offset voltage is perfectly cancelled, within the limit of the approximation made to obtain equation 4.13:

$$V_{O;Compensated} = 0 \quad (4.16)$$

If the offset voltage of the comparator $V_{OC} \neq 0$ is considered, the output of the comparator is:

$$D = \begin{cases} 1 & \delta + V_{OC} > 0 \\ 0 & \delta + V_{OC} \leq 0 \end{cases} \quad (4.17)$$

If the decision of the algorithm is still based on D, the offset cancellation is no more perfect and the final offset voltage becomes:

$$V_{O;Compensated} = V_{OC} \qquad (4.18)$$

This result indicates that in this circuit the performance of the offset compensation is limited by the offset of the comparator. This problem is due to the low level of the detection signal δ. It can be overcome by first compensating the offset of the comparator before using it to measure the offset of the amplifier. The compensation of the comparator is performed in the same manner as the amplifier calibration in the open-loop configuration which is presented below.

In the open-loop configuration, equation 4.14 shows that the offset can be efficiently sensed at the output of the amplifier, as illustrated in figure 71.

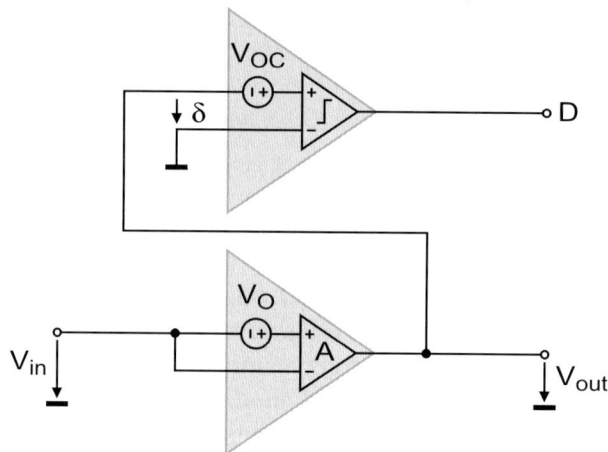

Figure 71. Offset measurement in the open-loop configuration

In this topology, the sensing is single-ended in the unique detection node which is the output of the operational amplifier. In the ideal case ($V_O = 0$), V_{out} is null and for this reason, the output voltage of the amplifier is compared to the ground voltage by the comparator. The detection signal δ is derived from equation 4.14:

$$\delta = AV_O \qquad (4.19)$$

Equation 4.17 still applies, but the digital circuitry must act conversely as in the closed-loop configuration. It decreases the offset if D = 1 ($\delta > 0$; $V_O >$

0) and increases it if $D = 0$ ($\delta \leq 0$; $V_O \leq 0$). This is because δ and V_O have the same sign, whereas they had opposed signs in equation 4.15.

Assuming an infinite resolution of the compensation, the final offset voltage is:

$$V_{O;Compensated} = -\frac{V_{OC}}{A} \qquad (4.20)$$

Table 12 summarizes the characteristics of both closed-loop and open-loop offset detections.

Table 12. Closed-loop and open-loop offset measurement

Parameter	Closed-loop	Open-loop
Detection signal δ	$-V_O$	AV_O
Compensated offset $V_{O;Compensated}$	V_{OC}	$-\dfrac{V_{OC}}{A}$

The open-loop configuration has two major advantages over the closed-loop topology. First, the signal level in the detection node is significantly higher, since it is multiplied by the open-loop gain A of the amplifier. The second advantage is a consequence of the first one. Since the signal level at the input of the detection circuit is higher, the effect of the imperfections of the latter (here the offset voltage V_{OC} of the comparator) is reduced. In the open-loop configuration, the final offset voltage after compensation is divided by the open-loop gain A of the amplifier. In both cases, the resulting offset is in fact the offset voltage of the comparator, referred to the input of the amplifier under compensation.

Based on the previous findings, it is possible to give the following guidelines for the choice of the detection node(s). If possible, one should choose a detection node (or a pair of detection nodes) so that:

1. The detection signal δ is *a function of the imperfection(s) to be compensated only* and does not depend on other parameters.
2. The *detection signal level δ is high* compared to the noise level and to the imperfections of the detection circuit.

Obviously, these rules are general and the choice is also dictated by circuit trade-offs (the choice of a detection configuration and detection node is not always trivial) or system constraints (like continuous-time operation).

Finally, it is noteworthy that the comparator does not necessarily need to be implemented like an operational amplifier. In the case of the open-loop configuration for example, the detection signal level is in some applications high enough to allow the direct use of the output voltage V_{out} as digital decision. In fact, the output of the amplifier saturates to one of the power supply rails if the open-loop gain A is sufficiently high. As long as the amplifier does not enter its linear region (when V_O is small) it behaves as a comparator. An intermediate solution between a complex comparator and no comparator at all is to use a digital buffer, as presented in figure 72.

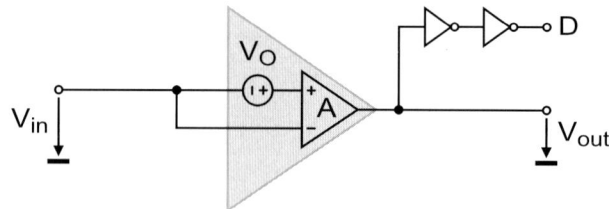

Figure 72. Implementation of a comparator with a digital buffer

In fact, the input/output characteristics of the CMOS inverter are those of a degraded inverting comparator, as shown in figure 73.

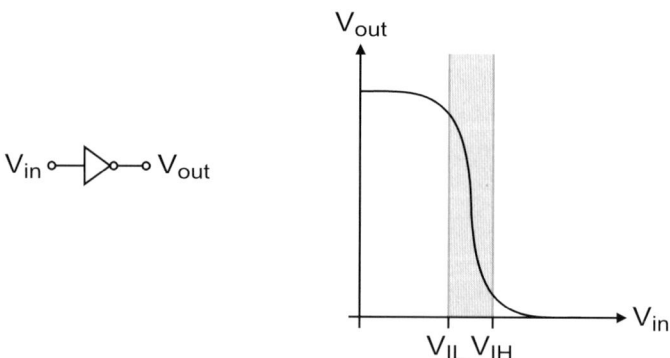

Figure 73. Input/output characteristics of the CMOS inverter

V_{IL} is the highest input voltage which is considered as a 0 logic level, whereas V_{IH} is the lowest acceptable input voltage for a 1 logic level. The shaded zone in the figure corresponds to the input voltage values for which

Chapter 4: Digital compensation of analog circuits 105

the output level of the inverter is undefined, i.e. where the output value is uncertain. Except in this range, the digital inverter behaves like a comparator. In the open-loop configuration of figure 72, the best-achievable offset correction is:

$$|V_{O;Compensated}| \leq \frac{V_{IH} - V_{IL}}{A} \qquad (4.21)$$

3.3 Compensation node

Based on the measurement made in the detection node, the digital algorithm corrects the imperfection by injecting a compensation current in the compensation node. The compensation can also be differential by using two complementary compensation nodes.

The compensation by means of the injection of a *correction current* is well suited for circuits based on MOS transistors. The devices themselves indeed basically transform a control voltage (the gate voltage V_G) into a current (the drain current I_D). The output impedance g_{ds} of the device, possibly combined with another impedance in parallel, converts I_D into a voltage. These general considerations apply to a large majority of circuits, principally amplifiers.

Since the signal is primarily conveyed by a current, and possibly further converted into a voltage by the output impedance, it is coherent to correct a possible imperfection of the device by adding a compensation current at the output. This situation is depicted in figure 74. The representation in this figure combines a small-signal representation and a small-signal transposition of a DC imperfection. This is done for explanation purpose only and should not be used in circuit modelling.

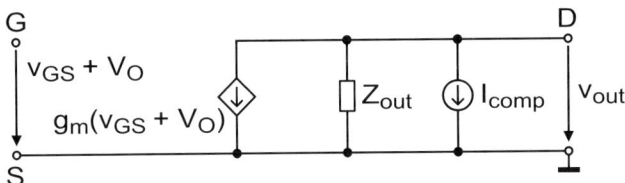

Figure 74. Compensation by current injection

The input signal v_{GS} is amplified by the transconductance g_m of the transistor, and transformed into a voltage by the output impedance Z_{out}. The bias current (not represented) is added to this differential signal, as well as an offset current I_{offset} which corresponds to the input offset voltage V_O of the device. In fact, I_{offset} is the difference between the real threshold voltage V_T

of the transistor and the ideal case, multiplied by the transconductance g_m at the bias point:

$$I_{offset} = g_m V_O \qquad (4.22)$$

In fact, it is usually not the absolute value of the V_T of one single transistor that is important, but the mismatch between the threshold voltages of two transistors. In this case, the difference can be minimized by using matching and increasing circuit area. But the simplified model presented above also applies.

Consequently, the offset current I_{offset} corresponds to the input offset voltage that has to be compensated. This can be done by injecting a compensation current I_{comp} into the output node (the drain D) to cancel I_{offset}:

$$I_{comp} = -I_{offset} \qquad (4.23)$$

The offset is cancelled and the transistor behaves exactly as if the V_T was ideal, i.e. the output voltage is:

$$v_{out} = Z_{out}[g_m(v_{GS} + V_O) + I_{comp}] = Z_{out} g_m v_{GS} \qquad (4.24)$$

It is noteworthy that these equations represent only the small signals, and that the DC bias point is not considered. Furthermore, the offset is included into this small-signal analysis, in spite of the fact that it is also a DC signal. This is done on purpose to show an approximation of the value of the offset current I_{offset} and consequently the compensation current I_{comp}. However, this representation is of no other interest in circuit modelling.

Let's analyze how the compensation current can be injected and how to choose the most appropriate compensation node based on the example of the Miller operational amplifier. Different approaches can be implemented. In [47], [48] and [49], a second differential pair is connected in parallel and the offset is corrected by applying a compensation voltage V_{comp}. Figure 75 presents such a circuit.

This topology has the disadvantage of being very sensitive to the voltage difference V_{comp}. In fact, the main differential pair is usually designed for high gain, and it is difficult to design a second differential pair with a much degraded gain using similar devices. The compensation voltage V_{comp} thus remains in the order of magnitude of the offset voltage to be compensated, typically a few millivolts. This reduced signal level degrades the signal-to-noise ratio and is not optimal. Another disadvantage of this technique is that it uses a voltage compensation signal. If the digital-to-analog converter used to generate the compensation signal has a current output, it is necessary to transform it to a voltage.

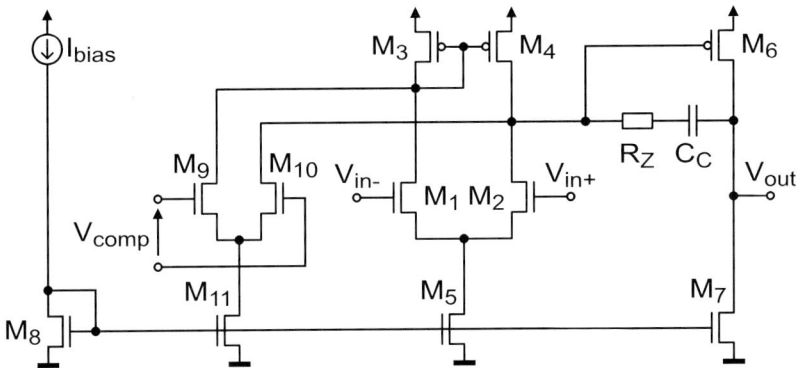

Figure 75. Offset correction by additional differential pair

A second approach is to modify the current mirror [50] that implements the active load of the differential pair [13]. One additional transistor is added in series with each branch of the current mirror. They are operated in the linear region and act as variable resistors controlled by their gate voltage. By applying a voltage difference V_{comp} to the gates of these transistors, the current mirror is unbalanced and the offset is compensated. Figure 76 shows how a degenerated active load is realized.

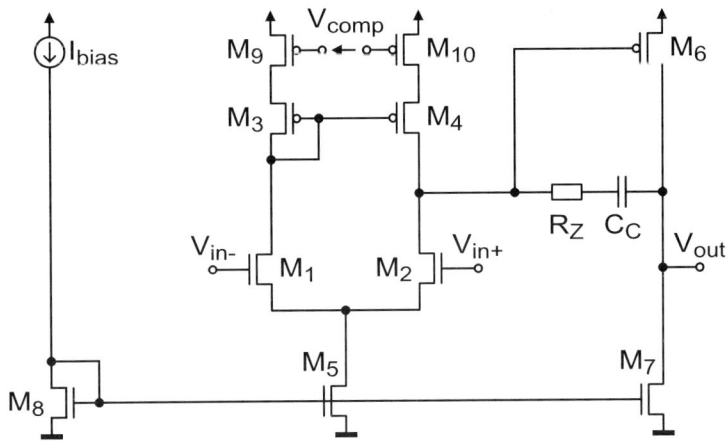

Figure 76. Offset correction by degenerated current mirror

This compensation circuit has the same disadvantage of being controlled by a voltage. Furthermore, the transistors in series with the branches of the current mirror cause an additional voltage drop that complicates the design with low supply voltages.

There is a more convenient way to compensate the offset of the Miller operational amplifier, by directly injecting a compensation current without current-to-voltage conversion. This approach is presented in [12] for the Miller amplifier and in [51] for a Differential Differences Amplifier (DDA). Figure 77 shows the circuit presented in [51].

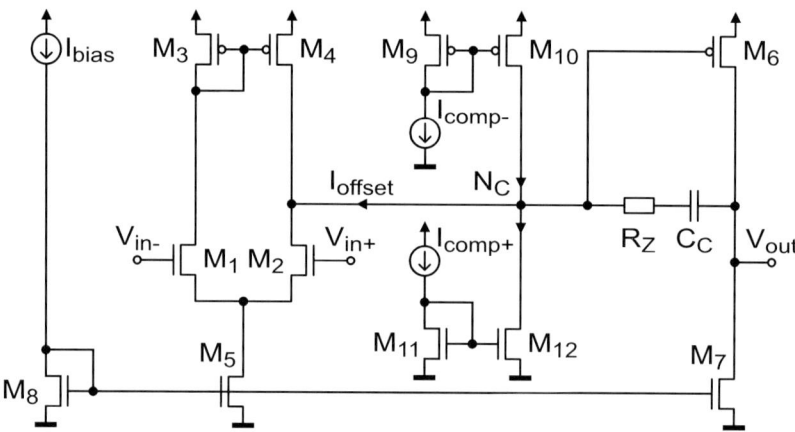

Figure 77. Offset correction by unilateral current injection

The four transistors M_9 to M_{12} inject two complementary compensation currents I_{comp+} and I_{comp-} in the compensation node N_C. The resulting compensation current is:

$$I_{comp} = I_{comp+} - I_{comp-} \qquad (4.25)$$

The input offset that has to be compensated, and thus the resulting offset current I_{offset}, can be positive or negative. In contrast, the sign of the current in the output transistor of the current mirror is fixed. For this reason, both current sources are necessary: M_{10} pushes current into N_C to compensate positive offsets, whereas M_{12} pulls current from N_C when the offset is negative.

The current sources I_{comp+} and I_{comp-} can basically be operated in two different manners. In single-ended mode, one of the current sources is kept constant, whereas the other current is generated by a digital-to-analog converter which is controlled by the digital compensation algorithm. In differential mode, both currents are the complementary outputs of a single DAC.

If the maximum input offset voltage is $V_{offset;max}$, the maximum offset current is:

$$I_{offset;max} = g_{m1} V_{offset;max} \tag{4.26}$$

The corresponding range for the compensation current is:

$$-I_{offset;max} \leq I_{comp} \leq I_{offset;max} \tag{4.27}$$

If the compensation is done in single-ended mode with a fixed I_{comp-} current, one must have:

$$I_{comp-} = I_{offset;max} \tag{4.28}$$

in order to generate the most negative compensation current. On the other hand, to cover the complete range of equation 4.27, the full scale of the DAC must be:

$$FS = 2 I_{offset;max} \tag{4.29}$$

In the differential mode, it is sufficient to ensure that both I_{comp+} and I_{comp-} generate currents in the [0, $I_{offset;max}$] range to satisfy equation 4.27:

$$0 \leq I_{comp-/+} \leq I_{offset;max} \tag{4.30}$$

A problem that can arise in the circuit of figure 77 because of the adjunction of M_{10} and M_{12} is a reduction of the gain in the first stage. In fact, it is equal to:

$$A_1 = \frac{g_{m1}}{g_{ds2} + g_{ds4} + g_{ds10} + g_{ds12}} \tag{4.31}$$

If the length L of transistors M_{10} and M_{12} is small and/or their current large, the term $g_{ds10} + g_{ds12}$ in equation 4.31 becomes dominant and decreases the gain A_1 in the first stage. This can be avoided by appropriately sizing these two transistors so that their channel conductance remains acceptable. Another solution is to use the alternative circuit of figure 78.

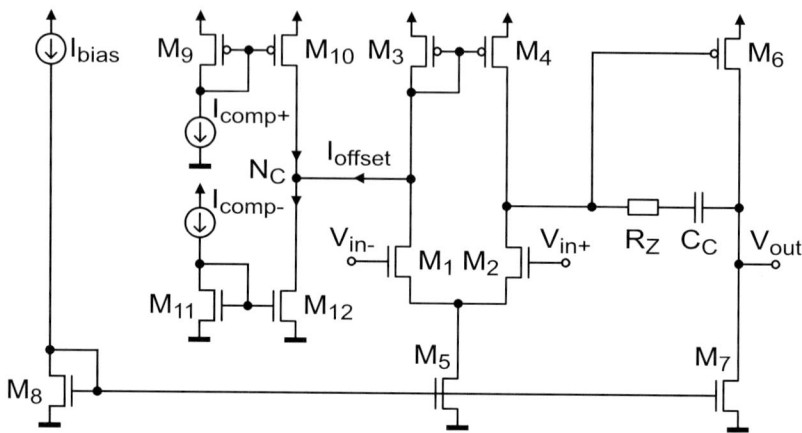

Figure 78. Offset correction by improved unilateral current injection

In this circuit, the impedance at the N_C node (without the compensation circuitry) is dominated by the diode-connected transistor M_3, and is approximated by:

$$Z_{N_C} = \frac{1}{g_{m3}} \quad (4.32)$$

This value is significantly smaller than in the previous case where it is equal to R_1 (equation 4.1), because $g_m > g_{ds}$. For this reason, the conductance added to N_C by M_{10} and M_{12} is negligible.

The other potential problem that must be considered is the additional parasitic capacitance due to M_{10} and M_{12}. The parasitic capacitance C_P in the compensation node is responsible, along with g_{m3}, for the third pole of the Miller amplifier:

$$p_3 = \frac{g_{m3}}{C_P} \quad (4.33)$$

It so happens that p_3 is compensated by a zero at twice this frequency. Furthermore, it is usually higher than the bandwidth of the amplifier and only slightly affects its stability [7].

Chapter 4: Digital compensation of analog circuits

If a differential compensation is considered, the circuit presented in figure 79 is well-suited.

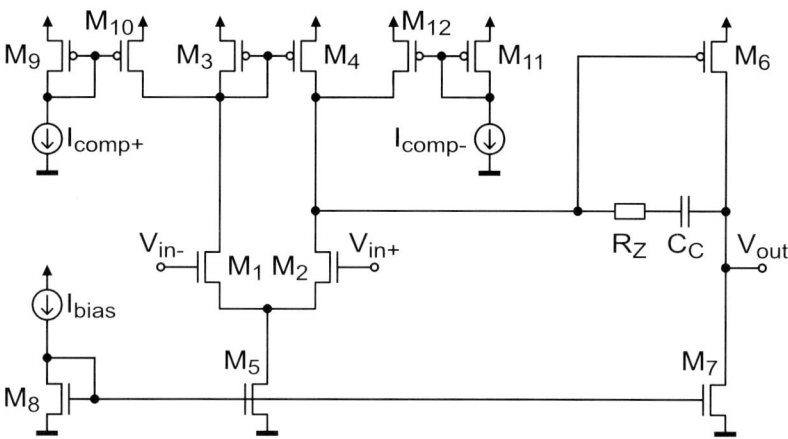

Figure 79. Offset correction by bilateral current injection

In this circuit, the current mirrors are located on both sides of the differential pair, contrary to the preceding ones (figures 77 and 78). Instead of having 2 complementary current mirrors (one NMOS and one PMOS), both mirrors are here PMOS and the current input is in the same direction. This allows the direct connection of an M/2$^+$M ladder to the compensation current inputs, transistors M_9 and M_{11} being the current collectors as in chapter 3, section 11.4.

The dimensioning of the compensation current mirrors in the case of the Miller operational amplifier is done as follows: Firstly, the maximum current must be determined by equations 4.28 and 4.29 or equation 4.30. Secondly, the transistor length L is determined in order to have a sufficiently low channel conductance g_{ds} that does not affect much the impedance of the compensation node. The length must also be sufficient to limit the effect of voltage variations in the compensation node on the compensation current that is injected. In saturation mode (and strong inversion), the drain current depends on the drain-to-source voltage V_{DS} (channel length modulation effect):

$$I_D = \frac{\mu C_{ox}}{2} \cdot \frac{W}{L}(V_{GS} - V_T)^2(1 + \lambda V_{DS}) \qquad (4.34)$$

Thirdly, the transistor width is calculated, considering that the output transistors of the current mirrors must remain in saturation. The saturation voltage of a MOS transistor is (in strong inversion):

$$V_{DS;sat} = \sqrt{\frac{2I_D}{\mu C_{ox}} \cdot \frac{L}{W}} \qquad (4.35)$$

The width must be large enough to keep the saturation voltage below the voltage in the compensation node.

It is noteworthy that the input and output currents of the current mirror do not need to be matched accurately. In fact, it is precisely the output current of the mirror that is injected into the compensation node, and thus adjusted by the successive approximations algorithm. The exact value of the collected current at the output of the ladder and its transformation by the mirror to the output current has no importance. For this reason, increasing the dimensions of the transistors composing the current mirror in order to improve their matching is superfluous and would degrade circuit performance by increasing the parasitic capacitance in the compensation node. For these reasons, the dimensions of the current mirrors injecting the compensation current can remain small.

Finally, concerning the choice of the compensation node(s), it is possible to give the guidelines set out below. If possible, one should choose a compensation node (or a pair of compensation nodes) so that:

1. The compensation current corrects an equivalent imperfection current corresponding to the *imperfection(s) to be compensated only* and does not have other effects on the circuit.
2. The voltage level and variation allow to design a small current mirror, considering the *channel length modulation* effect and the *saturation voltage*.
3. The influence of the addition of the current mirror on the *impedance* and the *parasitic capacitance* in the compensation node, as well as on *system parameters* linked to these values, is limited.

These rules may not apply to some particular cases. But in general, they allow the identification of the most suitable node.

To conclude, multiple effects can be compensated in the same node by a unique compensation current. For example, the equivalent offset of all the amplifiers in a amplification chain can be compensated in one single amplifier

[52]. Section 5 gives another example of single compensation of multiple imperfections in a SOI 1T DRAM.

3.4 DAC resolution

The necessary resolution of the DAC generating the compensation current can be calculated from the initial and final magnitude of the imperfection. In fact, it corresponds to the ratio between the worst-case value of the imperfection before correction, and the maximum allowable value after compensation.

As explained in section 3.3, there is a direct link between the input offset of the Miller operational amplifier and the equivalent offset current in the compensation node (equation 4.26). If the correction is perfect, the compensation current completely cancels the offset current and the input offset becomes null. But in the general case, the compensation current I_{comp} is not exactly equal to the offset current I_{offset} and there is a resulting offset current after compensation:

$$I_{offset;comp} = I_{offset} + I_{comp} \qquad (4.36)$$

If a successive approximations algorithm is used to perform the compensation, the maximum difference between the final value I_{comp} and the ideal value $-I_{offset}$ (equation 4.23) is smaller than the LSB b_1 of the DAC (equation 3.3), and thus:

$$\left|I_{offset;comp}\right| \leq b_1 \qquad (4.37)$$

Furthermore, the full-scale of the DAC should be chosen to cover the complete range of imperfection currents:

$$FS = \Delta I_{offset} = I_{offset;max} - I_{offset;min} \qquad (4.38)$$

Dividing equation 4.38 by 4.37 gives:

$$\frac{\Delta I_{offset}}{\left|I_{offset;comp}\right|} \geq \frac{FS}{b_1} \qquad (4.39)$$

The right term of inequation 4.39 is the resolution of the converter (equation 3.11), and thus:

$$\text{Res} \leq \frac{\Delta I_{offset}}{|I_{offset;comp}|} = \frac{g_m \Delta I_{offset}}{g_m |I_{offset;comp}|} = \frac{\Delta V_{offset}}{|V_{offset;comp}|} \quad (4.40)$$

The quality of the compensation, i.e. the ratio of the initial to final imperfection current (I_{offset}), is better than or equal to the resolution of the converter. As shown in equation 4.40, the necessary resolution can also be calculated directly from the ratio of the initial to final imperfection themselves (V_{offset}). The necessary resolution corresponds in fact to the worst-case reduction of the magnitude of the imperfection. If the offset has to be reduced by at least a factor of 16 for instance, 4 bits of resolution are necessary.

3.5 Low-pass decision filtering

Because the successive approximations algorithm is based on a series of decisions at fixed time intervals, it is sensitive to noise. The working condition of the successive approximations algorithm ensures that a rejected bit b_i can be (almost) compensated by the least significant bits (b_{i-1} down to b_1). But if the algorithm takes the erroneous decision of keeping a bit b_i because of an external interference, this implies that the output code of the DAC never comes back below this overestimated value.

During the compensation of the offset of an operational amplifier, the input noise can perturb the measurement at the instant of decision of the successive approximations algorithm. To avoid this, the mean value of the output of the operational amplifier should be considered for the decision, instead of the instantaneous value. The averaging can be done in the analog domain by inserting a low-pass filter between the output of the amplifier and the comparator [53], as presented in figure 80.

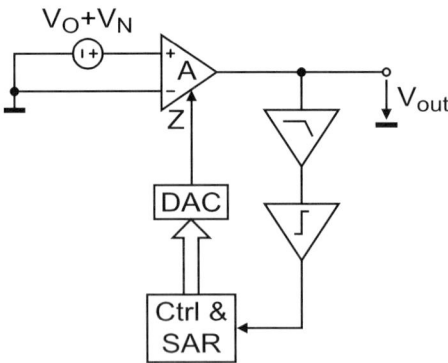

Figure 80. Analog averaging of the offset measurement

Chapter 4: Digital compensation of analog circuits   115

The low-pass filter attenuates the high-frequency noise voltage component V_N, and the decision of the algorithm is based on the continuous value V_O only. In fact, the pole of the low-pass filter should be located at the frequency of the compensation. In this way, the output of the filter contains only frequency components that are compensated by the auto-zero. In particular, the low-frequency components of the noise V_N are also cancelled, as explained in chapter 2, section 4.3.

The averaging can also be performed in the digital domain, by inserting an averager between the comparator and the digital control circuit implementing the successive approximations algorithm. Figure 81 shows the resulting circuit.

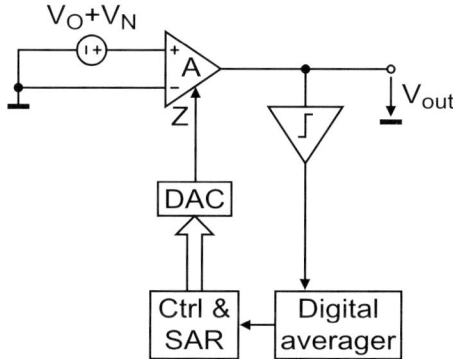

Figure 81. Digital averaging of the offset measurement

The digital average can be obtained by performing a sequence of observations of the digital output of the comparator at close time intervals, and by extracting the majority value. In this case, the majority calculation can be implemented simply by an up-down counter.

3.6 Continuous-time compensation

This section analyses the digital correction techniques that can be used when a continuous compensation is necessary to permanently eliminate a circuit imperfection that is time-varying. Two categories of systems are considered: the continuous-time systems, where the signal processing is performed continuously, and the sampled systems, where the signal is observed at fixed time intervals.

As explained in section 3.1, the detection of the imperfection is not always possible during normal circuit operation. In some cases, the circuit must be removed from the signal path and placed in a special detection configuration for this purpose.

In continuous-time systems, the removal from the signal path is not possible unless ping-pong is used to provide an alternative processing circuit during calibration.

In sampled systems, it is possible in some cases to place the circuit in a different detection configuration and to perform the calibration between two sampling intervals. If this is not feasible, the ping-pong with two structures being alternately calibrated can be used, without the drawback of glitch generation at the instant of exchanging circuits as in continuous-time systems.

Concerning the injection of the compensation current, the reasoning is the same as for the detection configuration. In sampled systems, it can be done using the successive approximations algorithm between two samples, or with ping-pong.

In continuous-time circuits without ping-pong, the injection of the compensation current must be done with care, avoiding fast and/or large variations. The successive approximations algorithm cannot be used directly, because during its execution, it produces abrupt changes of the compensation current, with amplitudes up to half the full scale. Figure 82 shows the typical look of the tracking of an imperfection using successive approximations. The imperfection is plotted in grey and the compensation in black. The algorithm performs 5 tests, one for each bit of the DAC, and when the calibration is done at t_{done}, it starts again.

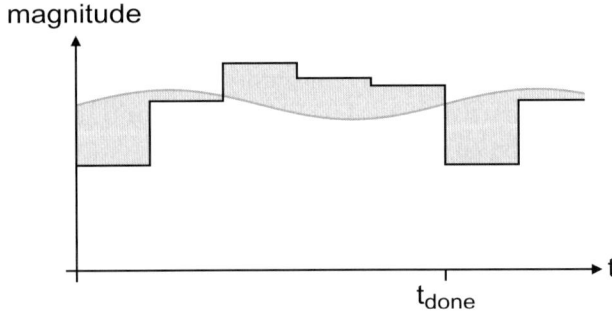

Figure 82. Imperfection tracking with successive approximations

The difference between the imperfection and its compensation is the shaded area. It is the remaining imperfection, which is much larger than the LSB of the compensation DAC in this case. This situation is not acceptable, because it creates temporary large imperfections. This is contrary to the purpose of the algorithm.

The solution is to reduce the magnitude of the steps generated by the DAC by using a modified compensation algorithm. With the same digital decision input, namely if the imperfection is higher or lower than its nominal value, it periodically increases or decreases the compensation value by only one LSB. The imperfection is continuously tracked by a digital compensation ramp. Figure 83 shows the behavior of this new compensation scheme on the same example as in figure 82. The imperfection and the LSB of the DAC are identical in both plots.

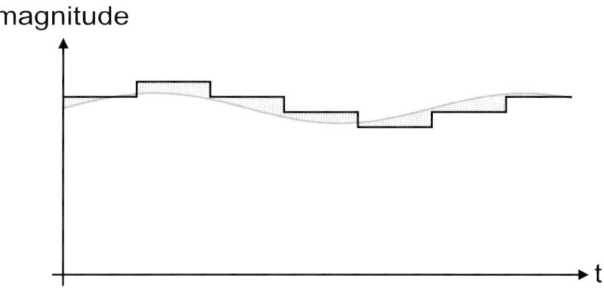

Figure 83. Imperfection tracking with up/down

A simple way to realize the digital control logic is an up/down counter. However, this implies that the digital output of this counter be connected to a linear DAC, and excludes the use of a sub-binary converter. If the necessary resolution of the converter is low, a M/2M structure can be considered. It is linear and for just a few bits of resolution, the area of the ladder remains acceptable. In [12], a 6 bits M/2M DAC is implemented using a unity transistor with an aspect ratio (W/L) of 3/3. If a higher resolution is necessary, two sub-binary converters controlled by a successive approximations algorithm and the up/down current mirror structure presented in the next section can be used.

3.7 Up/down DAC

The circuit presented in this section allows to generate ramps with sub-binary DACs. It uses the ping-pong technique between two converters: During the first phase, one converter generates the compensation current, whereas the other one is adjusted by a successive approximations algorithm to produce the next step, which is the same current increased or decreased by the step size ε. An intermediate phase allows the smooth switching between the two current sources. Then, the roles of the converters are exchanged and the second converter is adjusted, whereas the previously tuned one generates the compensation current. Figure 84 presents the functioning principle of the circuit.

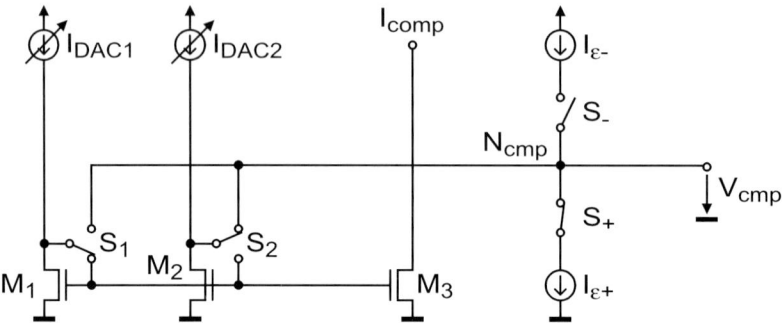

Figure 84. Up/down current mirror principle

The main component of the circuit is the current mirror with two inputs (M_1 and M_2) and one output (M_3). Each of the two sub-binary DACs is connected to one input, and the output is the compensation current I_{comp} that is injected in the compensation node. In figure 84, the switches S_1 and S_2 are configured so that the current I_{DAC1} is copied to the output by M_3, but also to the second input branch by M_2.

The two current sources $I_{\varepsilon+}$ and $I_{\varepsilon-}$ generate the up and down step currents respectively. Switch S_+ is closed (as shown in the figure) if the current must be increased, whereas S_- is closed if the next compensation current has to be smaller by ε than the present current $I_{comp} = I_{DAC1}$.

The voltage output V_{cmp} allows to adjust the current I_{DAC2} generated by the second converter in order to produce the next step value, smaller or larger by ε than I_{DAC1}. In the comparison node N_{cmp}, assuming that the output impedance of transistor M_2 and of the current sources I_{DAC1}, $I_{\varepsilon+}$ and $I_{\varepsilon-}$ are all very high, the voltage V_{cmp} clips either to the positive or negative power supply rail, depending on whether there is excess or lack of current. In other words, the comparison node N_{cmp} acts as a current comparator and converts the current difference into 2 extreme voltage levels, which is a digital information. A successive approximations algorithm is used to adjust I_{DAC2} in order to null the total current in N_{cmp}, by using V_{cmp} as decision input.

At the end of the algorithm, if the DAC has an infinite resolution, the total current in the comparison node is null and consequently:

$$I_{DAC2} = I_{DAC1} + I_{\varepsilon+} \qquad (4.41)$$

Conversely, if switch S_- is closed and S_+ is open:

Chapter 4: Digital compensation of analog circuits

$$I_{DAC2} = I_{DAC1} - I_{\varepsilon-} \quad (4.42)$$

Once this new compensation current is adjusted, the current mirror can perform a smooth transition by using both inputs simultaneously, as shown in figure 85.

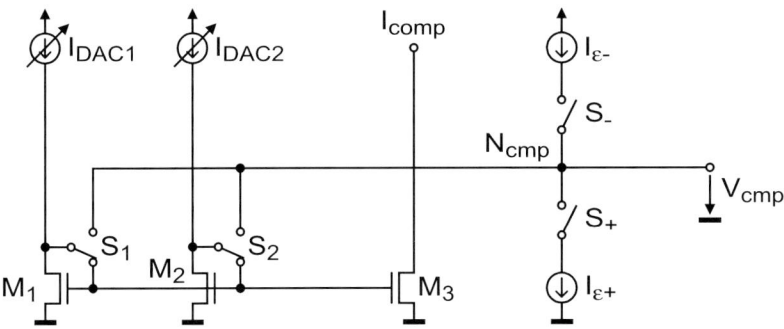

Figure 85. Smooth transition during up/down step

In this configuration, the output compensation current is:

$$I_{comp} = \frac{I_{DAC1} + I_{DAC2}}{2} \quad (4.43)$$

This allows to reduce by a factor of 2 the instantaneous variation of the compensation current, by dividing the transition in two half-amplitude steps. In the final configuration, the switch S_1 is toggled and I_{DAC1} is connected to the comparison node N_{cmp}. It can then in turn be adjusted to generate the next step.

It is shown below that the up/down current mirror is robust to V_T and aspect ratio mismatch between the transistors. The channel length modulation effect, which modifies the drain current as a function of the drain to source voltage V_{DS}, is not included in this analysis. In fact, the transistors of the current mirror are cascoded in the real implementation.

The drain current of a MOS transistor in saturation (and strong inversion), neglecting the channel length modulation effect (equation 4.34), is:

$$I_D = \frac{\mu C_{ox}}{2} \cdot \frac{W}{L}(V_{GS} - V_T)^2 \quad (4.44)$$

To simplify the notation:

$$\beta_i = \frac{\mu C_{ox}}{2} \cdot \frac{W_i}{L_i} \qquad (4.45)$$

And thus, the drain current of transistor M_i is:

$$I_{Di} = \beta_i (V_{GS} - V_{Ti})^2 \qquad (4.46)$$

The gate to source voltage V_{GS} is the same for the tree transistors. During the first phase, when I_{DAC1} is copied to the output and I_{DAC2} is adjusted, the output current is:

$$I_{comp} = I_{D3} = \beta_3 (V_{GS} - V_{T3})^2 \qquad (4.47)$$

At the end of the adjustment of I_{DAC2}, when the step is done by toggling both switches S_1 and S_2, the new drain current I'_{D2} of M_2 is:

$$I'_{D2} = I_{D2} + I_\varepsilon = \beta_2 (V'_{GS} - V_{T2})^2 \qquad (4.48)$$

Where V'_{GS} is the new gate to source voltage for the three transistors, and:

$$I_\varepsilon = \begin{cases} I_{\varepsilon+} & S_+ \text{ is closed} \\ -I_{\varepsilon-} & S_- \text{ is closed} \end{cases} \qquad (4.49)$$

V'_{GS} is extracted from equation 4.48:

$$V'_{GS} = V_{T2} + \sqrt{\frac{I_{D2} + I_\varepsilon}{\beta_2}} \qquad (4.50)$$

And V_{GS} from equation 4.46:

$$V_{GS} = V_{T2} + \sqrt{\frac{I_{D2}}{\beta_2}} \qquad (4.51)$$

The difference between the new and old gate voltage is:

Chapter 4: Digital compensation of analog circuits

$$\Delta V_{GS} = V'_{GS} - V_{GS} = \sqrt{\frac{I_{D2} + I_\varepsilon}{\beta_2}} - \sqrt{\frac{I_{D2}}{\beta_2}} = \sqrt{\frac{I_{D2}}{\beta_2}} \left(\sqrt{1 + \frac{I_\varepsilon}{I_{D2}}} - 1 \right) \qquad (4.52)$$

Equation 4.52 can be simplified by using a first-order Taylor development of the $\sqrt{1+x}$ function:

$$\Delta V_{GS} \cong \sqrt{\frac{I_{D2}}{\beta_2}} \left(1 + \frac{I_\varepsilon}{2I_{D2}} - 1 \right) = \frac{I_\varepsilon}{2\sqrt{\beta_2 I_{D2}}} \qquad (4.53)$$

The corresponding output compensation current variation is:

$$\begin{aligned} \Delta I_{comp} &= I'_{D3} - I_{D3} = \beta_3 [(V'_{GS} - V_{T3})^2 - (V_{GS} - V_{T3})^2] \\ &= \beta_3 [(\Delta V_{GS} + V_{GS} - V_{T3})^2 - (V_{GS} - V_{T3})^2] \\ &= \beta_3 \{\Delta V_{GS} [\Delta V_{GS} + 2(V_{GS} - V_{T3})]\} \end{aligned} \qquad (4.54)$$

If the internal term ΔV_{GS} is neglected, equation 4.54 becomes:

$$\Delta I_{comp} \cong 2\beta_3 \Delta V_{GS} (V_{GS} - V_{T3}) \qquad (4.55)$$

Finally, replacing equations 4.47 and 4.53 in 4.55 gives:

$$\Delta I_{comp} \cong 2\beta_3 \frac{I_\varepsilon}{2\sqrt{\beta_2 I_{D2}}} \sqrt{\frac{I_{D3}}{\beta_3}} = I_\varepsilon \frac{\sqrt{\beta_3 I_{D3}}}{\sqrt{\beta_2 I_{D2}}} = I_\varepsilon \frac{I_{D3}(V_{GS} - V_{T2})}{I_{D2}(V_{GS} - V_{T3})} \qquad (4.56)$$

This result shows that the step amplitude ΔI_{comp} of the compensation current is roughly equal to the step current I_ε. Two major effects are present in equation 4.56. First, the term I_{D3}/I_{D2} expresses that the step amplitude depends on the ratio of the absolute currents. If the aspect ratios W_2/L_2 and W_3/L_3 are different, the absolute currents are proportional (assuming equal V_T and V_{GS}) to:

$$\frac{I_{D3}}{I_{D2}} = \frac{W_3/L_3}{W_2/L_2} \qquad (4.57)$$

Consequently, it is normal that the difference ΔI_{comp} of the drain currents is proportional to this ratio. Furthermore, if there is a mismatch of the aspect ratios, it also affects the step amplitude.

The second effect, namely the mismatch of V_{T2} and V_{T3}, is expressed in equation 4.56 by the term $(V_{GS} - V_{T2})/(V_{GS} - V_{T3})$. This ratio also modifies the step amplitude.

It is noteworthy that the V_T or aspect ratio mismatches only slightly affect the step amplitude, and above all that they do not change the sign of I_ε, and thus the step direction (up or down). The up/down mirror can thus be realized even with poorly matched transistors, since the exact amplitude of the step is usually not important.

To calculate the step size during the second ping-pong phase, when I_{DAC1} is adjusted, all the preceding equations can be used by replacing the parameters of M_2 by those of M_1.

An important point for the implementation is the channel length modulation effect. If the drain to source voltage affects the drain current of the transistors composing the up/down mirror, the current adjustment is not carried out correctly and the system malfunctions. To solve this problem, the transistors composing the current mirror are cascoded.

Figure 86 shows a complete schematic of the up/down current mirror.

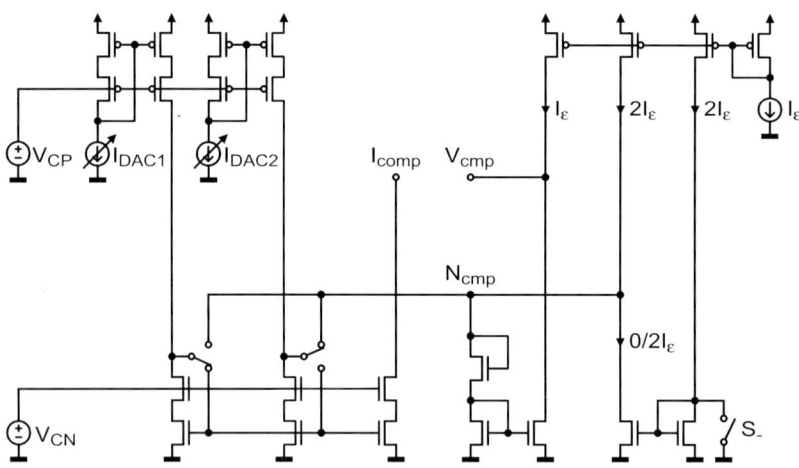

Figure 86. Up/down current mirror schematic

In the left part of the circuit, the current mirror presented in figure 84, as well as the input current sources, are cascoded. V_{CN} and V_{CP} are respectively the NMOS and PMOS cascode bias voltages.

The right part of the circuit of figure 86 implements the current comparator and the step up and step down currents. The decision voltage V_{cmp} is generated by a common source amplifier, which copies the resulting current of the comparison node N_{cmp} and transforms it into a voltage. The role of the diode-connected transistor that is inserted between the comparison node and the input of the current mirror is to create a voltage drop V_T in order to keep the voltage in N_{cmp} high enough. In fact, the transistors of the cascoded mirror should remain in saturation.

In the equilibrium state, the bias current of the comparator is equal to I_ε. The current generated by the up/down source is $2I_\varepsilon$ if the switch S_- is closed, or 0 (null) if S_- is open. In the cascode structure, the PMOS mirror pushes a current I_{DAC2}, whereas the NMOS mirror pulls I_{comp}.

Since the sum of all these currents in N_{cmp} is equal to the bias current I_ε of the common source amplifier in the equilibrium state, the adjusted current value I_{DAC2} of the DAC can be calculated. If the switch S_- is closed:

$$I_{DAC2} = I_{comp} - I_\varepsilon \tag{4.58}$$

If S_- is open:

$$I_{DAC2} = I_{comp} + I_\varepsilon \tag{4.59}$$

If a successive approximations algorithm bases its decisions on the voltage output V_{cmp} of the comparator to adjust I_{DAC2}, this circuit can be used as a ramp generator. The next compensation current step is up if S_- is open, and down if S_- is closed. In addition to the analog circuit of figure 86, two sub-binary DACs, a digital implementation of a successive approximations algorithm, and a digital control circuit are necessary to implement the up/down DAC.

The circuit has been extensively simulated using Monte Carlo analyses. The system has been integrated in a 0.8 µm process, and the complete up/down DAC tested. Simulations and measurements demonstrate the good functioning of the system. Figure 87 shows a micrograph of the up/down current mirror circuit of figure 86.

The area of the circuit is 0.015 mm^2 only, as the current mirrors do not need to be matched precisely. The philosophy of the up/down DAC is the same as for M/2$^+$M converters: a low circuit area and no need for precise components, which allows the implementation in modern digital technologies. Furthermore, the advantage of this digital solution is that it allows the indefinite conservation of the calibration current if no further up or down step is required.

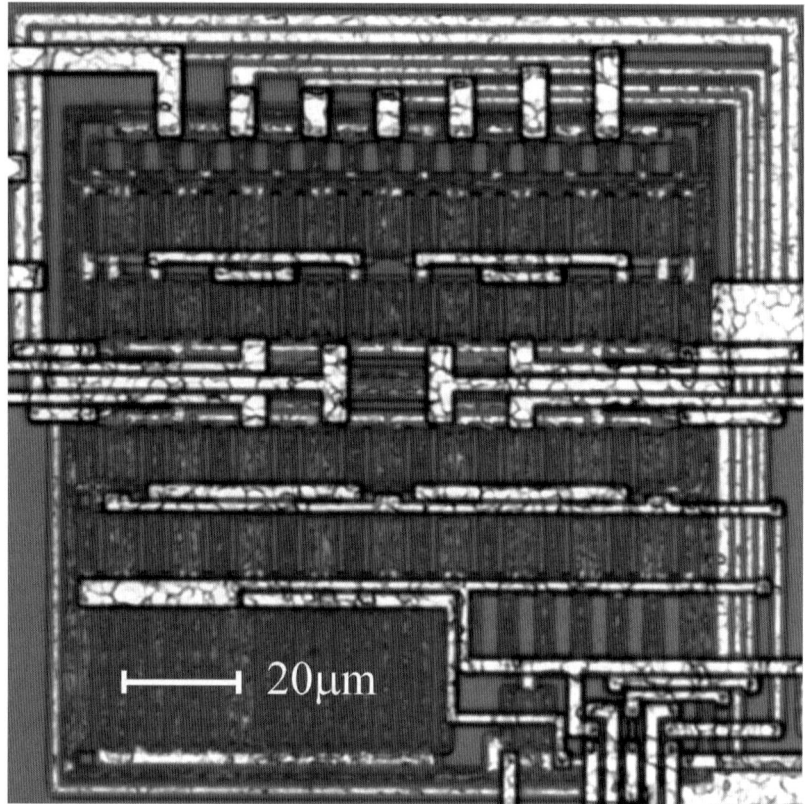

Figure 87. Up/down current mirror micrograph

4 SIMULATION WITH DIGITAL COMPENSATION CIRCUITS

The compensation methodology presented in this chapter allows a designer to improve the performance of an analog circuit by the addition of a digital correction circuit that injects a compensation current in an internal node. Once the designer has identified a detection configuration, a detection node and a compensation node, he can rely on the correction circuit to perform an automatic compensation.

During the design phase however, the commercial simulators do not offer complete simulation tools for supporting automatic calibration blocks. This section presents a simulator extension that allows the transparent simulation of circuits containing digital calibration blocks. All the common analyses can

Chapter 4: Digital compensation of analog circuits 125

be performed with and without the compensation block to determine the most suitable topology, as well as the performances of the correction circuit.

The Compensation Components tool is implemented to interface with a commercial simulator in order to add these new possibilities.

4.1 Principle

The extension of the simulator is composed of 2 principal parts. First, a new automatic *compensation component* is introduced in the simulation environment and is available to the designer for being inserted in schematics. Second, an external software, the *simulation manager*, is in charge of controlling the simulator and performing the 2-pass simulation (figure 88). The first simulation pass allows the calculation of the ideal compensation value, whereas the second pass is the simulation requested by the user.

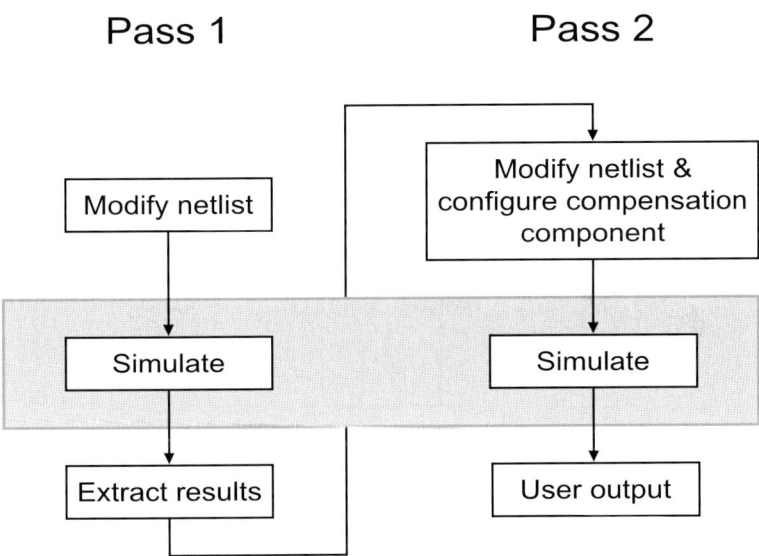

Figure 88. 2-pass simulation algorithm

At the beginning of the first pass, the simulation manager adapts the netlist and replaces the compensation component by a self-adjusting circuit (section 4.3). It then launches the simulator with the modified netlist. After the first pass, the simulation manager collects the necessary informations to configure the compensation component for the second pass. After adjusting the netlist a second time to compensate the circuit adequately with the configured compensation component (section 4.4), it then launches the simulation requested by the user.

The operation flow for the 2-pass simulation is performed by the simulation manager fully automatically and the adjustment of the compensation component is transparent. From the user point of view, the simulation of circuits using digital compensation circuits is thus exactly the same as conventional circuit simulation. Because the additional tool is simple and intuitive, the designer can focus on circuit optimization. In the analog circuit to be compensated, he can discover the ideal detection configuration, the detection and the compensation nodes. In the compensation circuit, he can determine the optimum resolution and current mirror (output stage) dimensions.

4.2 Automatic compensation component

The compensation components have 2 distinct behaviors. During the first simulation pass, they are replaced in the schematic by a circuit that allows the determination of the ideal compensation value. During the second pass, they are replaced by a configured block performing the adequate compensation, based on the results of the first pass.

In the schematic editor on the other hand, the compensation component has only one representation. This renders the compensation transparent for the user. Figure 89 presents the representation of a single-ended compensation component in the schematic editor.

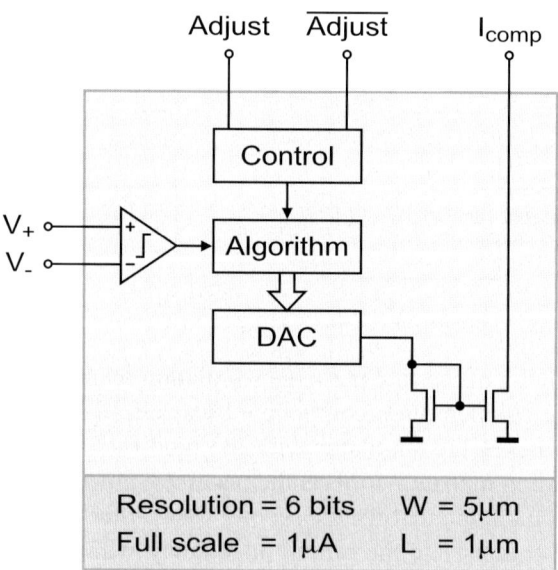

Figure 89. Single-ended compensation component in the schematic editor

Chapter 4: Digital compensation of analog circuits

The compensation component looks exactly like any other: It has inputs, outputs, a set of parameters and a symbolic representation. The automatic configuration feature is not apparent and is realized in the background.

The inputs V_+ and V_- are connected to the detection node(s) and used by the successive approximations algorithm to decide whether the compensation current I_{comp} must be increased or decreased. The I_{comp} output is connected to the compensation node.

The complementary digital output signals Adjust and $\overline{\text{Adjust}}$ are used to configure the circuit under compensation adequately in the detection configuration (when Adjust is active) or for normal operation (when Adjust is inactive).

The parameters of the compensation are the resolution and full scale of the DAC, and the dimensions W/L of the transistors composing the output current mirror.

Figure 90 presents the representation of the differential compensation component in the schematic editor. It has the same parameters, input and outputs as the single-ended component, except the current output that is duplicated and injects current in the compensation nodes in differential mode.

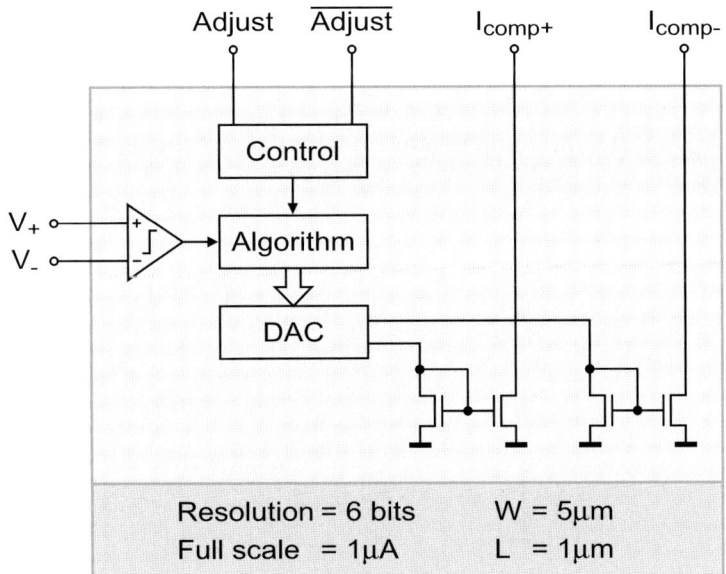

Figure 90. Differential compensation component in the schematic editor

4.3 Compensation component during adjustment

During the first simulation pass, the simulation manager modifies the netlist by replacing the single-ended compensation component by the circuit presented in figure 91.

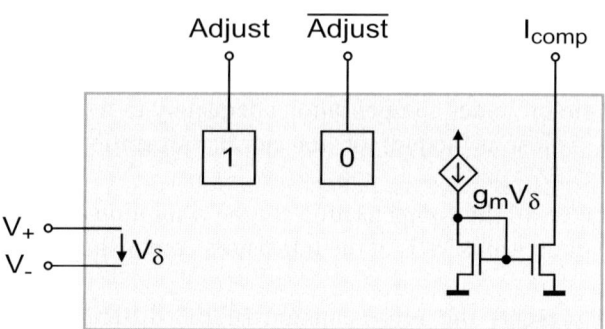

Figure 91. Single-ended compensation component netlist for the first pass

By applying a high logic level (1) on the Adjust signal and a low logic level (0) on $\overline{\text{Adjust}}$, the compensation component places the circuit in the detection configuration. The voltage difference V_δ between the sensing inputs V_+ and V_- is the detection signal, which controls the compensation current generated by the transconductance source. If the detection and compensation nodes are adequate, the compensation system closes an analog feedback loop that can be modeled by the circuit of figure 92.

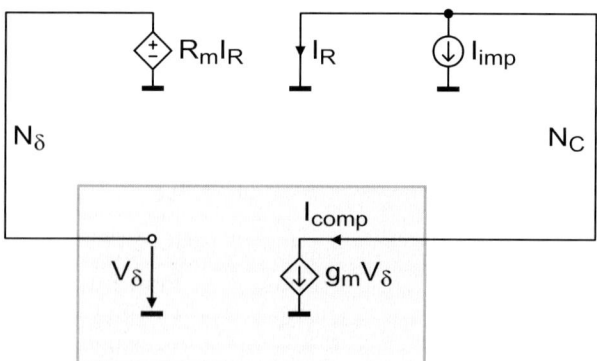

Figure 92. Model of the analog feedback loop of the first pass

The shaded block represents the compensation circuit, whereas the upper part models the circuit under compensation. In the compensation node N_C, the

Chapter 4: Digital compensation of analog circuits 129

source I_{imp} models the imperfection to be compensated. In the example of the offset of the Miller amplifier, this source is I_{offset}. This imperfection current is cancelled by the compensation current generated by the transconductance source g_m. If the cancellation is not perfect, there is a residue current I_R in the circuit that generates a detection signal voltage V_δ that can be observed in the detection node N_δ. This current to voltage transformation is modeled here by the transresistance R_m.

In the compensation node, the sum of the currents is null:

$$I_{imp} + I_{comp} + I_R = 0 \qquad (4.60)$$

with

$$I_{comp} = g_m V_\delta = g_m R_m I_R \qquad (4.61)$$

Extracting I_R from equation 4.61 and replacing it in equation 4.60 gives, after simplification:

$$I_{comp} = -\frac{1}{1 + \frac{1}{g_m R_m}} I_{imp} \qquad (4.62)$$

Which can be approximated, if the term $g_m R_m$ is large enough, by:

$$I_{comp} \cong -I_{imp} \qquad (4.63)$$

The compensation current thus almost completely cancels the imperfection current, and consequently the imperfection itself. The gain of the loop indeed depends on the transconductance g_m, which is thus voluntarily set to be extremely high in order to find the ideal compensation value with enough precision.

To obtain the ideal value of I_{comp}, the simulator is launched during the first pass with the modified schematic presented in figure 91, to compute only the operating point. This simple simulation is sufficient to find the optimal compensation value and only requires a short simulation time. In some cases, when the feedback loop includes sampled circuits in the system under compensation, the operating point simulation is replaced by a transient simulation. In these situations, the simulation time is increased. In all cases, the simulation manager extracts the ideal value from the output file of the simulator.

It is worth stressing that the *digital* adjustment performed by the successive approximations algorithm in the real circuit is done in the simulator by an *analog* feedback loop. This simplification saves simulation time, because an operating point analysis in this case allows to gather the same informations as the time-consuming transient simulation that extensively simulates the execution of the algorithm.

In the differential mode, the netlist used during the first-pass simulation is shown in figure 93.

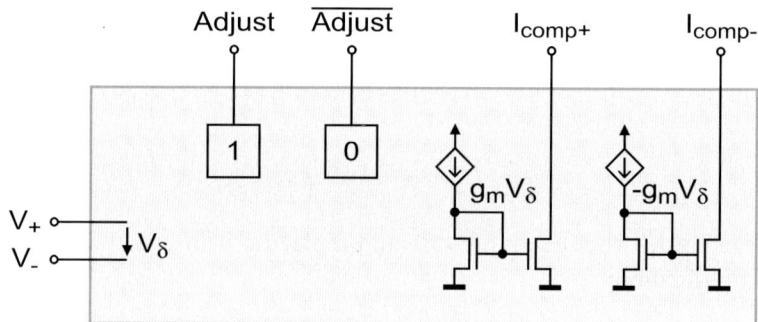

Figure 93. Differential compensation component netlist for the first pass

The output current is made differential by setting opposite signs to the transconductance sources.

4.4 Compensation component during compensation

During the first pass, the ideal value of the compensation current is determined by simulation of the circuit in the detection configuration. This information can then be used during the second pass, where the simulation analyses requested by the user are performed on the correctly compensated circuit. The compensation circuit that is used during the second pass for the single-ended mode is shown in figure 94.

Chapter 4: Digital compensation of analog circuits 131

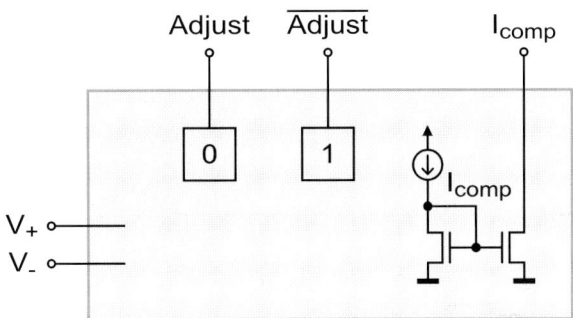

Figure 94. Single-ended compensation component netlist for the second pass

By applying a low logic level (0) on the Adjust signal and a high logic level (1) on $\overline{\text{Adjust}}$, the compensation component places the circuit in the normal configuration. The sensing inputs V_+ and V_- are not used, since the compensation current I_{comp} is fixed during this second pass by the simulation manager in function of the value $-I_{imp}$ extracted during the first pass.

The value set for I_{comp} by the simulation manager is chosen by the user between 4 possible values presented in table 13. Note that I_{imp} is a negative value, and thus $-I_{imp}$ is positive.

Table 13. Compensation currents for worst-case and Monte Carlo

I_{comp}	Comment
$-I_{imp}$	Ideal value
$-I_{imp} - \text{LSB}$	Worst-case value for conventional successive approximations
$-I_{imp} + \text{LSB}$	Worst-case value for reverse successive approximations
$-I_{imp} + \varepsilon$	Statistically distributed value

These 4 values correspond to the ideal, worst-case and statistical values of the conventional and reverse successive approximations algorithm, as depicted in figure 95.

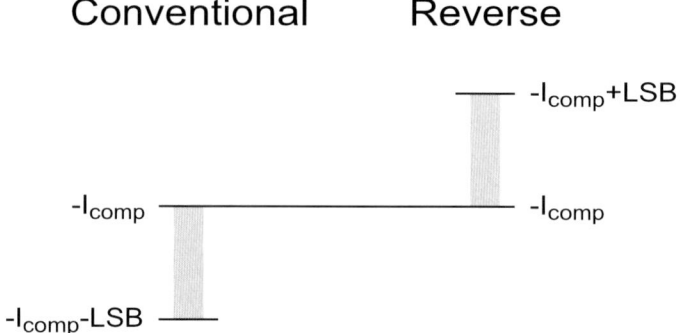

Figure 95. Final value range of the successive approximations algorithm
Left: Conventional algorithm; *Right*: Reverse algorithm

If the ideal compensation value is desired, the value $-I_{imp}$ extracted from the first pass simulation is used directly. This option can be chosen to verify the efficiency of the compensation in the best case, and to discover system parameters that degrade the compensation quality. For instance, if other parameters than the imperfection generate a detection signal, or if the current injection has unwanted secondary effects.

The worst-case compensation value depends on the type of the algorithm. For the conventional successive approximations algorithm, it is found from equations 3.3 and 3.4 to be 1 LSB smaller than the ideal value. For the reverse successive approximations algorithm, equations 3.3 and 3.8 indicate that the worst case is 1 LSB larger than the ideal value. For both algorithms, the other extreme value is the ideal value itself (perfect compensation). The LSB is computed from the DAC full scale and resolution, which are parameters of the compensation component.

The fourth possible choice for the compensation current is a randomly distributed value in the interval corresponding to the successive approximations algorithm or its reverse variant. For the conventional algorithm, a uniform distribution in the $[-I_{imp} - LSB, -I_{imp}]$ interval is used, with $\varepsilon \in [-LSB, 0]$. For the reverse algorithm, the value is randomly chosen in the $[-I_{imp}, -I_{imp} + LSB]$ interval, with $\varepsilon \in [0, LSB]$. If this statistical value of the compensation current is chosen, the complete circuit, including the compensation, is accurately modeled for Monte Carlo simulations.

In all four cases, the simulation manager limits the calculated compensation current to the full scale specified for the corresponding compensation component.

It is noteworthy that during the second simulation pass, the compensation current source is also modeled accurately, since it is the real output transistor of the current mirror that is connected to the compensation node.

Figure 96 shows the compensation circuit that is used during the second pass for the differential mode compensation.

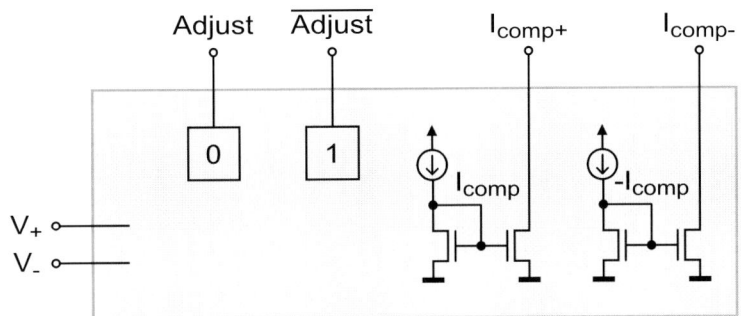

Figure 96. Differential compensation component netlist for the second pass

4.5 Multiple digital compensation

In some complex analog circuits, several digital compensation blocks may coexist, each one compensating one different imperfection. The methodology described in the previous section can be extended to this case.

If the circuits to be compensated do not depend on each other, they can be compensated simultaneously. By extension, the calculation of the ideal value of each compensation current can be done by one single simulation, as described in section 4.3. Except that several compensation values must be extracted instead of an unique one, no further modification is necessary.

If on the other hand, all the compensations cannot be done concurrently, the 2-pass simulation principle presented in section 4.1 must be adapted. A simple solution consists in adding a *sequence number* to each compensation component, which identifies the order in which the compensation components perform their respective calibration. It is then possible to perform the modified 2-pass simulation (figure 97). In the flow chart, the abbreviation CC stands for compensation component.

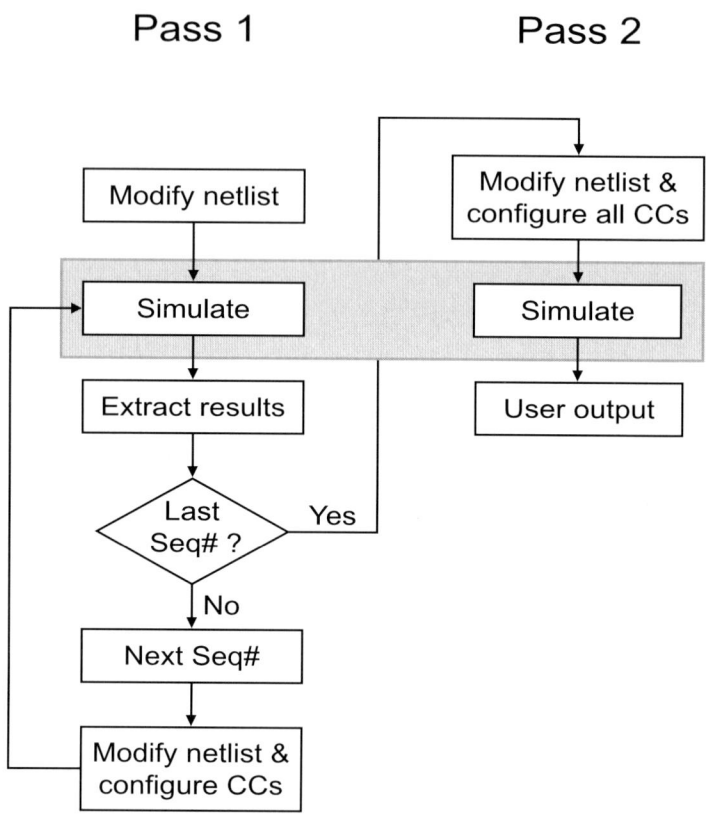

Figure 97. Modified 2-pass simulation algorithm

As many first passes are carried out as there are different sequence numbers. For some sequence numbers, there may be several simultaneous calibrations. At each step, the already calibrated compensation components, which have a smaller sequence number than the current one, are put in their compensation mode, as if they were already in the second simulation pass. Finally, when all the components have gone through the first pass, the simulation requested by the user is performed during the second pass.

4.6 Example of implementation for PSpice

The simulation methodology presented in this chapter is fully implemented for the PSpice simulator. The PSpice Device Equations extension allows to modify the built-in model equations, by adding modules developed in C++ and containing custom user-defined models.

Chapter 4: Digital compensation of analog circuits 135

Because these software extensions are closely linked to the core of the simulator, they allow the determination of the some internal simulation parameters, like the simulation run number in multiple simulations or Monte Carlo analyses. Furthermore, they benefit from the powerful possibilities of the C++ programming language, allowing for instance their interfacing with an external software.

Whilst these features are not necessary to implement the simulation methodology described in this chapter if one single simulation is performed, they are almost indispensable to allow multiple analyses, like Monte Carlo simulations. They allow to solve the problem of the different current values that need to be generated by the same component for each simulation run.

Short of giving implementation details, this section briefly shows which features the simulator must have to fully implement the simulation methodology. With some minor modifications, the presented technique can be adapted to almost all commercial simulators.

The diode is one of the components that can be modeled in the PSpice Device Equations extension. The conventional diode model is presented in figure 98.

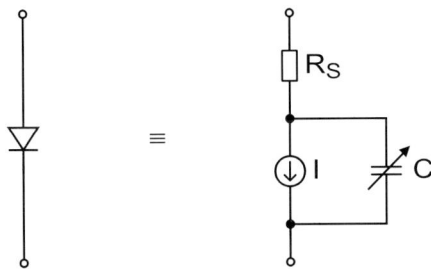

Figure 98. PSpice diode model

This diode model can be reduced to an ideal current source by setting $R_S = 0$ and $C = 0$. With a small overhead in the internal structure of the Device Equations extension module, it is possible to make the value of the current source dependent on the *simulation run number*, by reading the value for each run in a configuration file created during the first simulation pass. The resulting programmable current source is presented in figure 99.

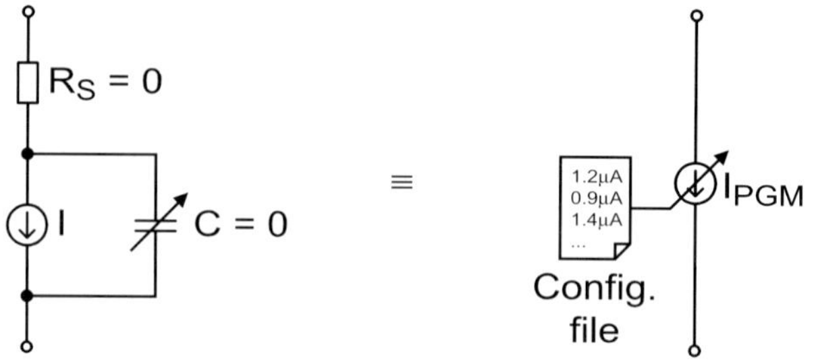

Figure 99. Programmable current source

Using this technique, a compensation component can appear perfectly compensated (or voluntarily deviated from the ideal value using another option from table 13) for each successive run of a Monte Carlo analysis. The condition is that the random seed for the Monte Carlo analysis is the same during the first and second pass.

The same result can be obtained without any modification of the simulator engine if only one simulation is performed at a time. The purpose of this section is to show the potential of a solution including multiple run simulations.

The complete simulation methodology presented in this chapter allows the accurate and complete simulation of digitally compensated analog circuits. All the conventional analyses can be performed without requiring complex manipulations, thanks to the simulation manager which performs transparent tuning of the digital blocks. The compensation circuits are modeled accurately and the influence of their design parameters can be simulated, including by worst-case and Monte Carlo analyses. By allowing easy, complete and accurate simulation of the circuits with compensation blocks, this methodology helps the designer to find the best way to improve the performance of analog circuits using digital compensation.

4.7   Offset compensation of the Miller amplifier

This section briefly presents the results of Monte Carlo simulations of the offset of the Miller operational amplifier. In particular, it shows the improvement obtained by using a digital compensation circuit and the effect of the resolution of the DAC on the performance. The simulations where realized using the PSpice extension presented in section 4.6.

Figure 100 presents the statistical offset dispersion of a typical untrimmed Miller amplifier, simulated on 1000 samples.

Chapter 4: Digital compensation of analog circuits 137

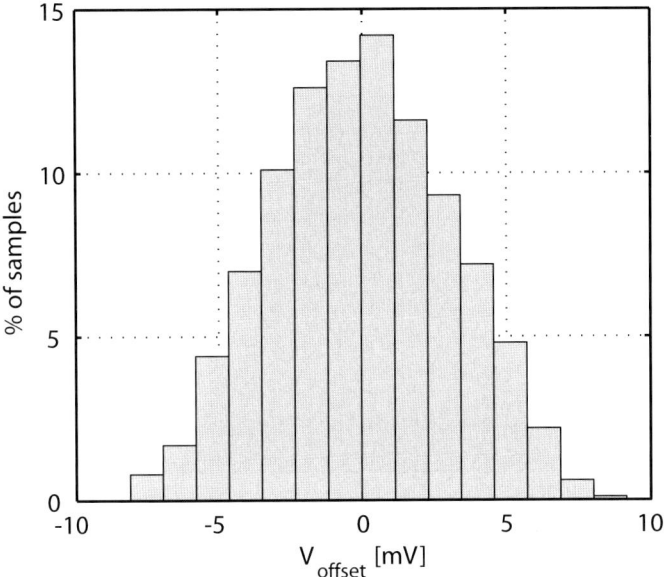

Figure 100. Untrimmed offset of a typical Miller amplifier

The mean offset value μ is -29 μV, and the standard deviation σ is about 3.1 mV. All the offsets are contained in the [-8.1, 9.2] mV interval.

The almost null μ indicates that the amplifier is carefully designed to avoid systematic offset. On the other hand, the statistical dispersion introduced by transistor mismatches, although it is a typical value for this kind of circuits, may be too high for some applications.

The simulation results for the same circuit, with a single-ended digital compensation based on a DAC with 8 bits of resolution, is presented in figure 101.

The new mean offset value μ is -33 μV, and the standard deviation σ is 20 μV. All the offsets are contained in the [-72.2, -0.2] μV interval.

It may appear strange at first sight that the gaussian offset distribution of the uncompensated amplifier becomes a uniform distribution. However, it is explained by the fact that ideal compensation current falls in a 1 LSB interval between the two nearest DAC output values, and that the probability of each value in this interval is equal.

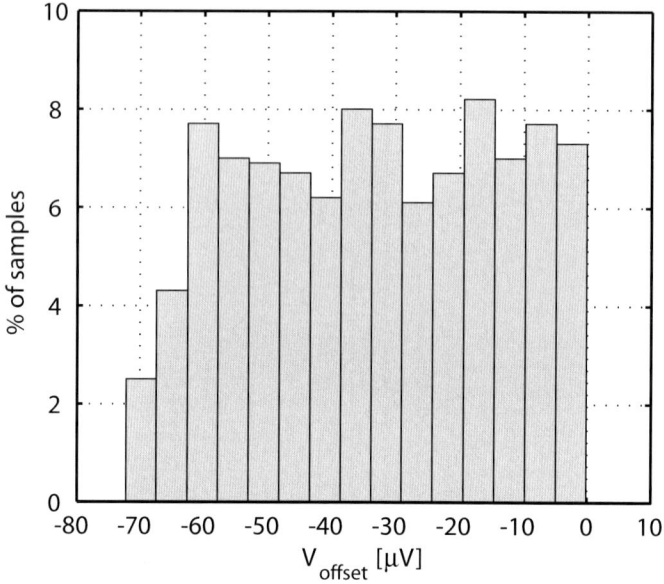

Figure 101. Miller amplifier offset with single-ended 8-bits trimming

The second observation is that the compensated offsets are always negative. This comes from the successive approximations algorithm, which finds the closest compensation current that remains smaller than the ideal value. Using reverse successive approximations would produce all positive offsets.

Finally, the offset reduction is the expected value for a compensation with 8 bits of resolution. The ratio between the worst-case values without and with compensation is:

$$\frac{\Delta V_{offset}}{\Delta V_{offset;comp}} \cong 240 \qquad (4.64)$$

This result is in close agreement with the prediction of equation 4.40, since for a resolution of 8 bits, one should have a factor of $2^8 = 256$ improvement.

5 APPLICATION TO SOI 1T DRAM CALIBRATION

This section presents the application of the digital compensation methodology detailed in the two last chapters to the automatic reference generation [54] for a Silicon On Insulator (SOI) memory circuit using single transistors

Chapter 4: Digital compensation of analog circuits 139

as memory cells (hence the name "1T DRAM"). This example demonstrates that several imperfections can, in some cases, be cancelled by one single compensation. Furthermore, it proves that the methodology is robust and can be exploited with low-quality analog components, even those manufactured using a fully-digital process.

5.1   1-transistor SOI memory cell

The 1T DRAM memory cell is realized in Partially Depleted (PD) SOI. It benefits from the fact that in SOI, the body of the transistors can be used to store charges, since it is insulated from the substrate. Figure 102 shows a cross section of a SOI transistor memorizing a 0 or 1 digital information by storing negative or positive charges, respectively, in its body.

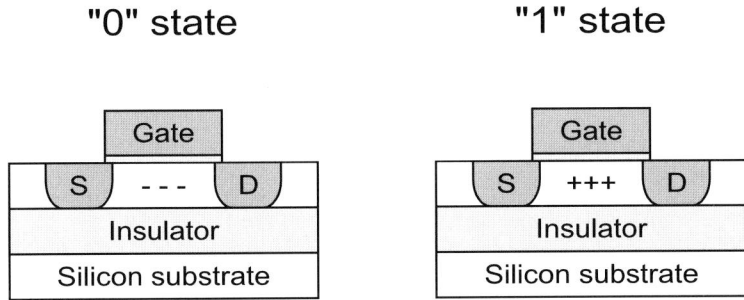

Figure 102.   SOI 1T DRAM cell
Left: Storage of state "0"; *Right*: Storage of state "1"

The information is written in the cell by applying adequate voltage levels to the terminals. The positive charge is created using impact ionization, whereas the negative charge is obtained by hole removal [55].

The charge stored in the body modifies the threshold voltage V_T and thus the drain current under determined bias conditions, allowing to read the memorized information. For an NMOS transistor, the resulting read current I_0 for a 0 information is lower than the drain current I_1 corresponding to a 1. By comparing the read current to an appropriate reference current I_{ref} in the $[I_0, I_1]$ interval, the stored information can be retrieved:

$$I_0 < I_{ref} < I_1 \tag{4.65}$$

5.2 Memory cell imperfections

When the same reference current is used for several memory cells, its value must be chosen carefully. In fact, there is a relatively important statistical dispersion of the drain currents for both states [56]. Figure 103 represents the typical statistical distribution of the read currents for both states.

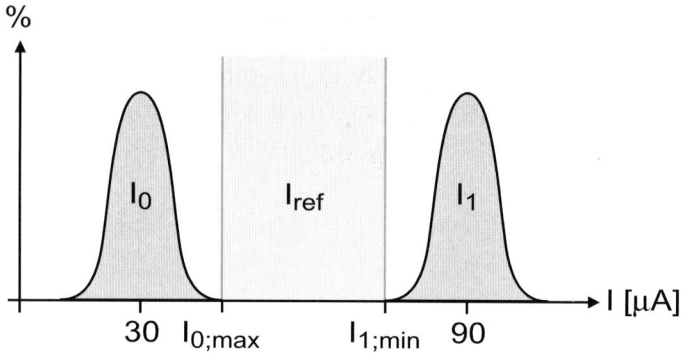

Figure 103. Read current dispersion of the 1T DRAM cell

The reference current window is the light shaded zone in the figure. To allow the correct reading of both states for all memory cells, the reference current must be chosen carefully between the largest possible current for state 0 ($I_{0;max}$) and the smallest current of state 1 ($I_{1;min}$). Equation 4.65 thus becomes:

$$I_{0;max} < I_{ref} < I_{1;min} \tag{4.66}$$

Another problem is the retention characteristics for both states. Whereas the current for a memorized 1 remains constant with time, I_0 progressively degrades and increases until reaching I_1 [57]. The typical retention time is in the order of 1 second. Figure 104 presents the typical read currents I_0 and I_1 as a function of time, and shows how I_0 degrades when no refresh cycle is performed.

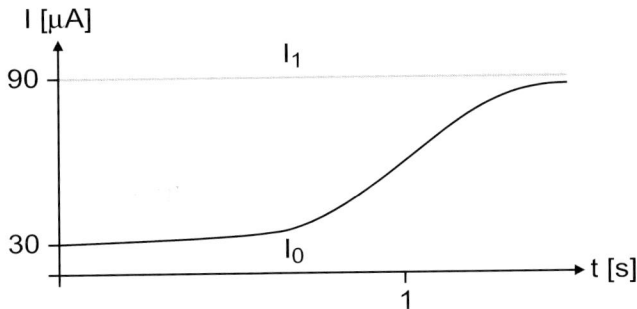

Figure 104. Retention characteristics of the 1T DRAM cell

In a memory circuit using these 1T DRAM cells, it is thus advisable to use a current reference that is near $I_{1;min}$ in order to increase the retention time. This is because the current window in fact shrinks, tending to $I_{1;min}$. In figure 105, the valid reference current values in function of time are located in the shaded area.

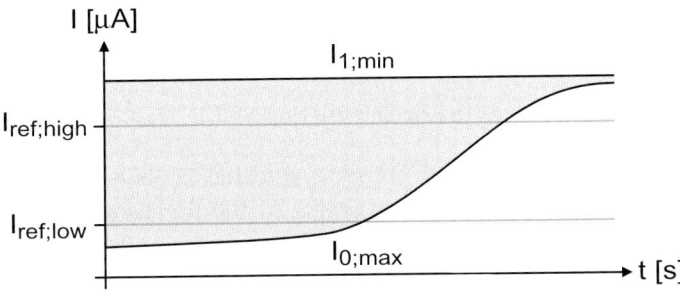

Figure 105. Reference current window as a function of time

If a low reference current $I_{ref;low}$ is used, it exits the current window sooner than if a higher value $I_{ref;high}$ is chosen. With $I_{ref;high}$, valid informations can thus be read from the memory cells longer without refresh (the retention time is longer).

5.3 Sensing scheme

In addition to the current reference, a current comparator (sense amplifier) is necessary to read from the memory cells. In conventional DRAMs, cross-coupled inverters are commonly used. A modified sense amplifier, specifically adapted to the SOI 1T DRAM, is proposed in [56]. Figure 106 shows the schematic of the sense amplifier, along with the two complementary bitlines

BL and $\overline{BL}$ where the memory cells MC and the reference currents are connected. Since memory cells are placed on both BL and $\overline{BL}$, two reference currents (one per bitline) are necessary. I_{ref} is the reference for the memory cells on BL, whereas $\overline{I_{ref}}$ is for the cells on $\overline{BL}$. Each memory cell in the memory circuit is connected at the intersection of a bitline BL and a wordline WL, which are respectively the rows and the columns of the memory matrix.

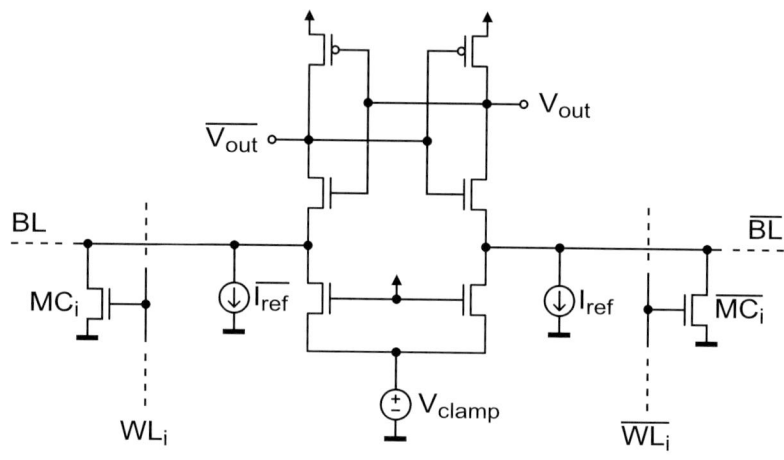

Figure 106. Sense amplifier for SOI 1T DRAM

The inputs of the sense amplifier are connected to the bitlines. In standard sense amplifiers, the inputs are the drains of the transistors composing the inverters. Here, they are located on the sources of the NMOS transistors. A voltage source V_{clamp} is connected to the bit lines through resistors implemented by two additional transistors. This modified architecture is necessary to operate the SOI memory cells with the specific voltage levels [57] necessary to perform the read cycle.

A series of switches (not represented in the figure) and control signals implement the appropriate read cycle timing. The memory cells are activated separately by their corresponding wordlines (WL), which are shared in the memory matrix between transistors connected to different bitlines. The result of the comparison of the memory cell current with the reference current is available at the two complementary outputs V_{out} and $\overline{V_{out}}$ of the sense amplifier.

A simplified model of the circuit of figure 106 is shown in figure 107. It represents the unilateral case with a memory cell on BL and the reference current on $\overline{BL}$. In the reverse situation, the analysis is similar.

Chapter 4: Digital compensation of analog circuits

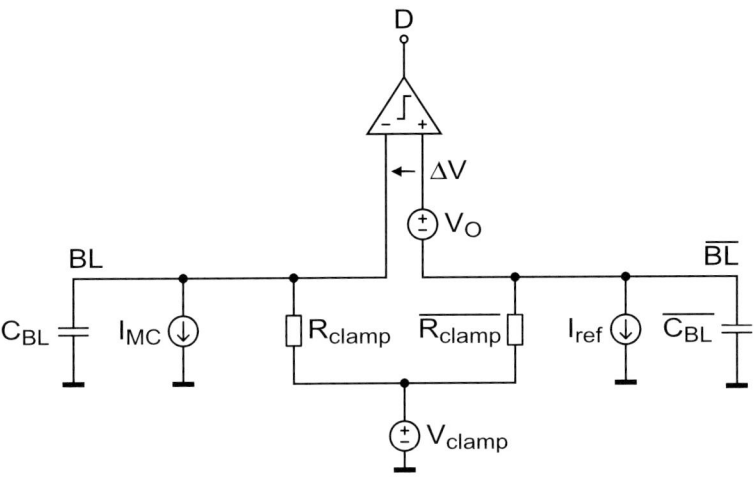

Figure 107. Sense amplifier model

The impedance of the clamping resistors is dominant and the parasitic bit-line capacitances C_{BL} and $\overline{C_{BL}}$ can be neglected. The R_{clamp} resistors thus convert the memory cell and reference current into voltages, which are compared by the sense amplifier. The offset voltage of the sense amplifier is modeled by the V_{offset} source.

The voltages in both input nodes of the comparator is:

$$V_- = V_{clamp} - R_{clamp} I_{MC} \qquad (4.67)$$

and

$$V_+ = V_{clamp} - \overline{R_{clamp}} I_{ref} + V_O \qquad (4.68)$$

The voltage difference ΔV between the two inputs of the ideal comparator is:

$$\Delta V = V_+ - V_- = V_O + R_{clamp} I_{MC} - \overline{R_{clamp}} I_{ref} \qquad (4.69)$$

The output D of the sense amplifier is 1 if the input voltage ΔV of the ideal comparator is positive ($\Delta V > 0$), and 0 otherwise ($\Delta V \leq 0$). The toggle point is at $\Delta V = 0$, which implies:

$$I_{ref} = \frac{V_O + R_{clamp}I_{MC}}{R_{clamp}} \qquad (4.70)$$

The adapted reference current depends on the memory cell current, but also on the matching between both clamp resistors and the offset of the sense amplifier. The reference current window is:

$$I_{ref} \in \left[\frac{V_O + R_{clamp}I_{0;max}}{R_{clamp}}, \frac{V_O + R_{clamp}I_{1;min}}{R_{clamp}} \right] \qquad (4.71)$$

5.4 Calibration principle

The ideal reference current I_{ref} can be found by a successive approximations algorithm. It compensates for all the imperfections of equation 4.70 and finds the highest value in the current window of equation 4.71 in order to maximize the retention time (see figure 105).

If the output D of the sense amplifier is used as decision signal in the successive approximations algorithm, the final adjusted reference value with a DAC having an infinite resolution is the value calculated by equation 4.70. If furthermore the current I_{MC} is the current $I_{1;i}$ generated by the memory cell MC_i storing a 1 data, the adjusted value $I_{ref;i}$ is:

$$I_{ref;i} = \frac{V_O + R_{clamp}I_{1;i}}{R_{clamp}} \qquad (4.72)$$

If this successive approximations algorithm is repeated for all the n memory cells MC_1 to MC_n on the bitline, and that the smallest value among the adjustment results $I_{ref;1}$ to $I_{ref;n}$ is taken, the result is:

$$I_{ref} = \min(I_{ref;1}, ..., I_{ref;n}) = \frac{V_O + R_{clamp}I_{1;min}}{R_{clamp}} \qquad (4.73)$$

This value is the upper limit of the current window of equation 4.71. If the DAC has a finite resolution, the use of the successive approximations guarantees that the final value is neither larger than the ideal value, nor smaller by more than 1 LSB of the converter (equations 3.3 and 3.4). In this case, equation 4.73 becomes:

Chapter 4: Digital compensation of analog circuits

$$I_{ref} = \frac{V_O + R_{clamp}I_{1;min}}{R_{clamp}} - \varepsilon \qquad (4.74)$$

with $\varepsilon \in [0, LSB]$. This is the highest value that the DAC can generate in the current window of equation 4.71. The automatic reference adjustment algorithm (figure 108) is used to obtain the ideal reference value of equation 4.74.

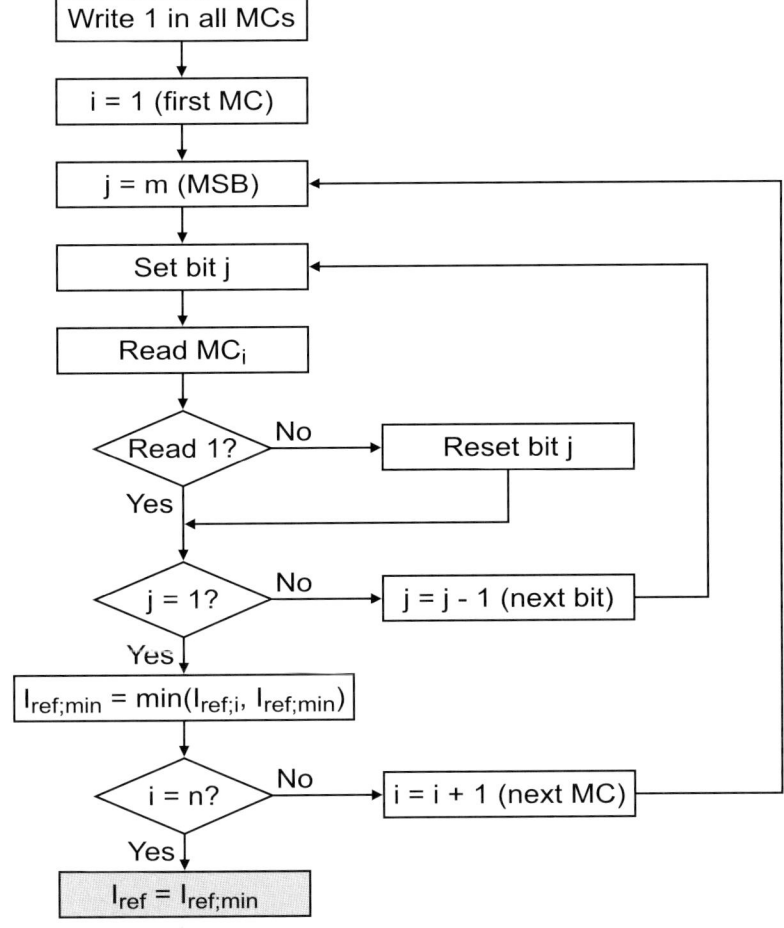

Figure 108. Automatic reference adjustment algorithm

First, the 1 data is written in all the memory cells. For each memory cell MC_i (outer loop), the successive approximations algorithm (inner loop) performs normal read cycles to determine which bits j of the DAC (m total bits) have to be kept. The adjusted current value is then compared to the global minimum $I_{ref;min}$, which is corrected if necessary. At the end of the algorithm, the reference current I_{ref} is set to $I_{ref;min}$, which is equal to the reference value calculated using equation 4.74. Note that the read operation is a simple and adequate test to allow the successive approximations to decide whether to keep or reject a given bit. In fact, an excessive reference value causes a 1 information to be read as a 0.

5.5   Calibration algorithm

The implementation of the automatic reference adjustment algorithm (figure 108) is simple. In particular, the minimum calculation can be done implicitly [58], using the optimized reference adjustment algorithm (figure 109).

The successive approximations algorithm is used to find a reference current for the first memory cell. It is then tested on the next cell. If the read cycle produces a correct output, the reference current is also suitable for this cell and the same test is performed again with the next cell. If on the contrary, the read cycle does not produce a correct 1 output for the new cell but a 0 instead, this means that the previously used reference is too high. The reference current needed for the new cell is smaller than the previous one and is the new minimum of the algorithm of figure 108. The successive approximations algorithm is used to find the new lower reference current.

The algorithm of figure 109 circumvents the minimum calculation by an analog current comparator. In the worst case, it performs n times (the number of memory cells connected to the bitline) the successive approximations algorithm. The worst-case total number of read cycles is thus in the order of $O(mn)$, where m is the number of bits of the DAC.

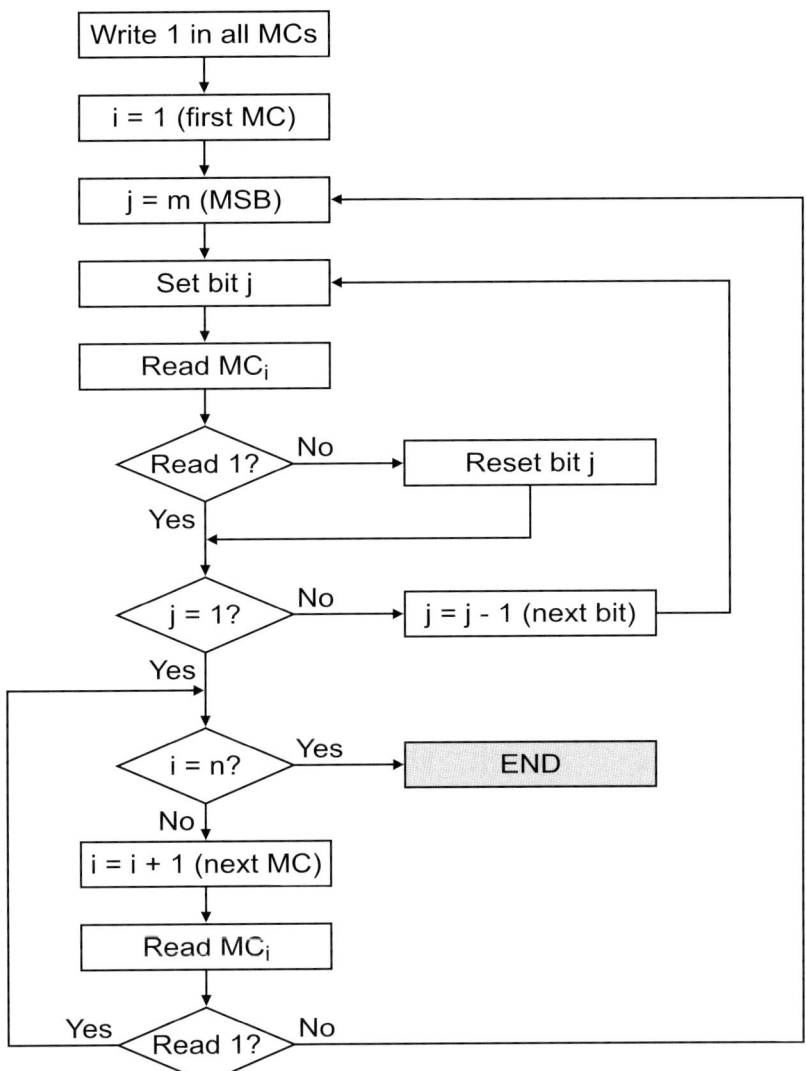

Figure 109. Optimized automatic reference adjustment algorithm

5.6 Measurements

The successive approximations algorithm presented in section 5.5 compensates for the mismatch of the clamp resistors and the offset of the sense amplifier. In addition, it adapts to the statistical dispersion of the currents generated by the memory cells. These imperfections are cancelled by one single detection and compensation, using the circuit itself to detect the imperfection.

A complete memory has been implemented [59] and tested. The measurements validate the calibration principle. Figure 110 shows the two complementary outputs of the sense amplifier during pattern reading in 3 adjacent memory cells. Successively, all 8 different combinations (000, 001, ..., 110, 111) are written and immediately read 3 consecutive times.

The results demonstrate that the current reference is properly adjusted, since all patterns are retrieved correctly, and that the read cycle is not destructive, as the same pattern can be read several times.

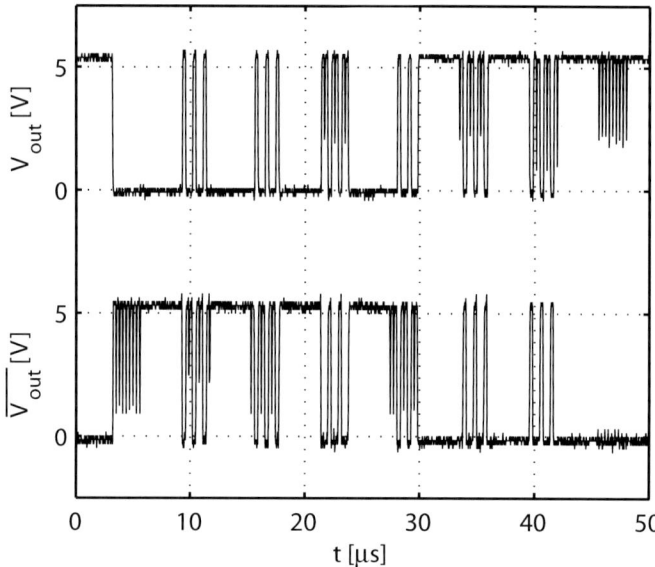

Figure 110. Write/read cycles on 3 adjacent memory cells

6 CONCLUSION

The sub-binary converters and the successive approximations algorithm presented in chapter 3 can be used to compensate a large variety of circuit imperfections. Using the compensation methodology presented in this chapter, the correction is done systematically. The simulation of digitally compensated analog circuits is possible thanks to a small extension of conventional simulators. This extension transparently adjusts the compensation blocks. It allows to analyze the efficiency of the correction, to determine the effects of the current injection on other circuit parameters and to design the most adapted digital compensation.

The methodology presented in chapters 3 and 4 allows to improve the performances of analog circuits by using digital compensation. It is especially interesting for fabrication technologies where the quality of the analog components is poor. By relaxing the constraints in analog design, it will allow in the future to design high-performance circuits using manufacturing processes which do not intrinsically provide sufficient component quality. By their design simplicity and their versatility, the digital correction circuits are fully adapted to the evolution of the technology and will support analog systems to overcome the new design challenges.

Chapter 5

Hall microsystem with continuous digital gain calibration

This chapter presents a digital gain calibration technique for Hall sensors. The gain is continuously measured using an integrated reference coil, without interrupting normal circuit operation. Appropriate modulation and demodulation schemes are implemented to separate the external and calibration magnetic fields. The gain drift is compensated by a digital compensation circuit. After an introduction and presentation of the state of the art, this chapter presents the system architecture, and chapter 6 its implementation.

1 INTRODUCTION

The gain drift of integrated Hall sensors is one of their current main limitations. The drift is due to temperature variations, mechanical stresses and ageing. The technique presented in this chapter allows a continuous cancellation of the gain drift using a real-time digital calibration system. The detection and compensation circuit uses the digital correction techniques and methodology presented in the previous chapters. The proposed system allows a gain drift reduction by a factor of 6 to 10 compared to current commercial products.

2 INTEGRATED HALL SENSORS

Different techniques and corresponding sensor technologies allow the measurement of magnetic fields [60]. Among them, the Hall sensor [61][62][63] is one of the most widespread solid-state sensors. It can be realized without requiring additional fabrication steps in commercial CMOS technologies, allowing the design of single-chip microsystems [64] comprising both the sensor [65] and its analog front-end. Mixed-mode solutions further including an analog-to-digital converter even allow the complete integration of smart sensors [66], combining a sensor and a dedicated digital signal processing circuit.

2.1 Hall effect

The Hall effect was discovered in 1879 by E. H. Hall [67]. It is a manifestation of the Lorentz force, which affects electrical charges in movement. For an electrical charge q having a vectorial velocity $\vec{v}$, the Lorentz force is:

$$\vec{F_L} = q\vec{v} \times \vec{B} \tag{5.1}$$

In conductors or semi-conductors, the carriers of the electrical current are also submitted to the Lorentz force. Figure 111 presents the case of a current I carried by positive charges q (e.g. holes in a semi-conductor) and flowing through a slab along the y axis between the current contacts C_+ and C_-. A perpendicular magnetic field $\vec{B}$ is applied along the z axis.

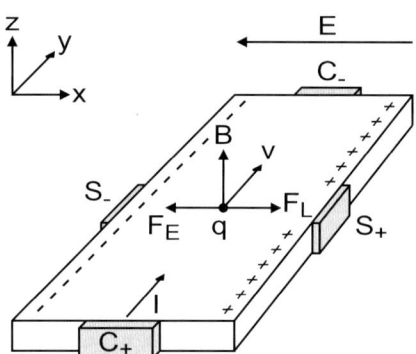

Figure 111. Hall effect

The Lorentz force deviates the positive carriers to the right of the slab and causes an opposite negative charge on the left. This charge accumulation creates a voltage difference along the x axis between the sense contacts S_+ and S_-, and an associated electrical field $\vec{E}$, which also affects the charges:

$$\vec{F_E} = q\vec{E} \tag{5.2}$$

Since there is no transversal current (along the x axis), the equilibrium state is reached when the force due to the electrical field (equation 5.2) cancels the Lorentz force (equation 5.1):

Chapter 5: Hall microsystem with continuous calibration   153

$$F_{L;x} + F_{E;x} = qv_yB_z + qE_x = 0 \qquad (5.3)$$

The voltage difference between S_+ and S_- is proportional to the magnetic field and is called Hall voltage.

2.2 Hall sensors

Hall sensors can be realized in CMOS technologies, without needing additional fabrication process steps. Various geometries can be implemented [68][63]. However, the cross-like shape presented in figure 112 is optimal [69].

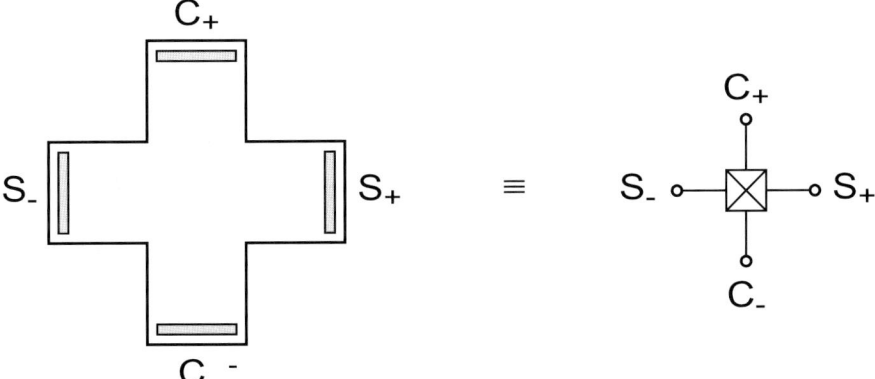

Figure 112. Cross-like Hall sensor and symbol

The generic electrical symbol of the Hall sensor is shown on the right-hand side of the figure. In a CMOS P-substrate N-well fabrication process, the cross-like Hall sensor can be implemented as presented in figure 113 [70], which shows a cross section of this sensor.

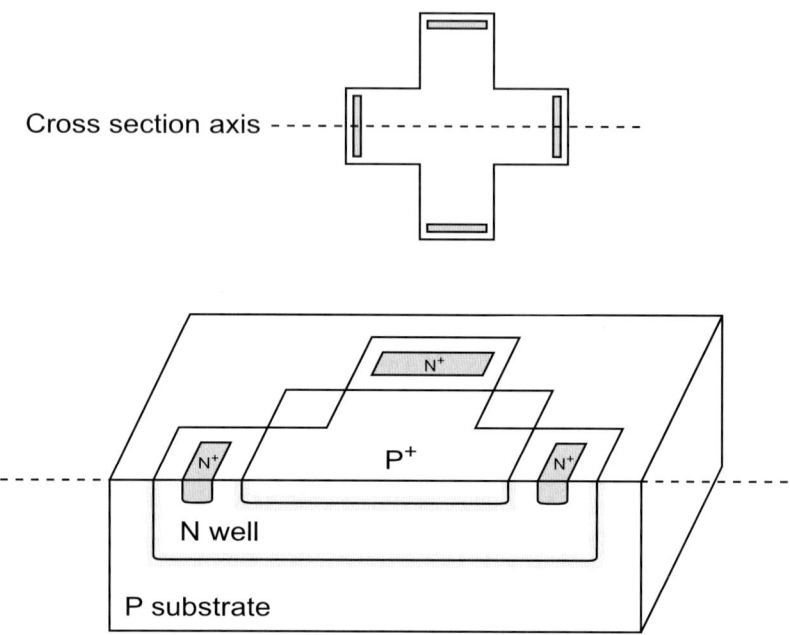

Figure 113. Cross-like Hall sensor implementation in P-substrate CMOS

The active part of the sensor is located in the N-well, in the "channel" between the two depletion layers (light grey) created at the interface of the well with the P substrate at the bottom and the P^+ shielding implant at the top. The 4 contacts are implemented using N^+ implants. Detailed explanations about the sensor design, optimization, functioning and simulation can be found in [63] and [69]. It is here simply accepted that the Hall voltage can be calculated as:

$$V_H = S_I I_{bias} B \qquad (5.4)$$

where I_{bias} is the sensor bias current, B the perpendicular magnetic field and S_I the current-related sensitivity, which is equal to:

$$S_I = G \frac{r_H}{qnt} \qquad (5.5)$$

In this equation, r_H is the Hall factor, q the elementary carrier charge, n the carrier density, t the thickness of the sensor and G a geometrical correction factor having a value in the [0, 1] interval, depending on the dimensions of the sensor.

If the sensor is biased by a voltage instead of a current, a similar relation to equation 5.4 can be derived:

$$V_H = S_V V_{bias} B \qquad (5.6)$$

where V_{bias} is the sensor bias voltage, and S_V the voltage-related sensitivity, calculated as:

$$S_V = G \frac{w}{l} \mu_H \qquad (5.7)$$

w and l are the width and length of the sensor respectively, and μ_H the carrier Hall mobility.

Integrated Hall sensors are subject to two main imperfections: the offset voltage and the drift of the sensitivity (S_I or S_V) due to temperature variations, mechanical constraints and ageing. While the offset voltage can be eliminated using the spinning current technique (section 3), the gain drift can be compensated using one of the techniques of section 4 or the continuous digital calibration technique presented at the end of this chapter.

2.3 Hall sensor models

Most models of Hall sensors are based on a Wheatstone bridge. The simplest one is the purely resistive bridge presented in figure 114, which models the passive behavior of the Hall sensor.

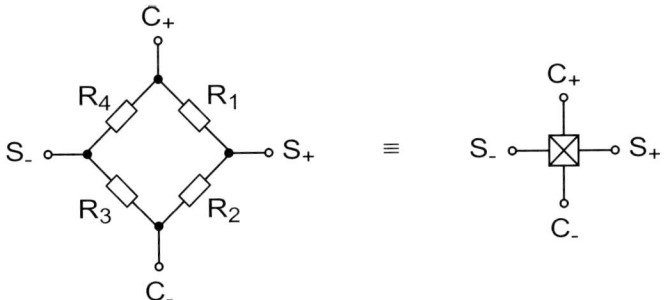

Figure 114. Purely resistive Hall sensor model

If the Hall element is symmetrical, as the cross-like sensor of figure 112, the values of the 4 resistors R_1, R_2, R_3 and R_4 composing the bridge are nominally identical to value R. However, due to fabrication process variations and the piezo-resistive effect [69], an imbalance is introduced. This mismatch causes the offset of the sensor and can be modeled by an additional ΔR resistor which is added or subtracted in each branch, as indicated in figure 115.

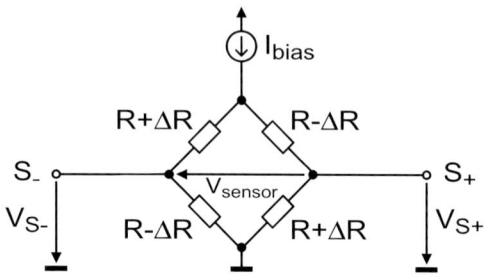

Figure 115. Modelling of the offset of the Hall sensor

When a bias current I_{bias} is applied to the sensor, a sensor offset voltage V_{OS} builds up between the sensing terminals:

$$V_{sensor} = V_{S+} - V_{S-} = V_{OS} = \Delta R I_{bias} \tag{5.8}$$

Figure 116 presents an extension of the previous model adding the Hall effect.

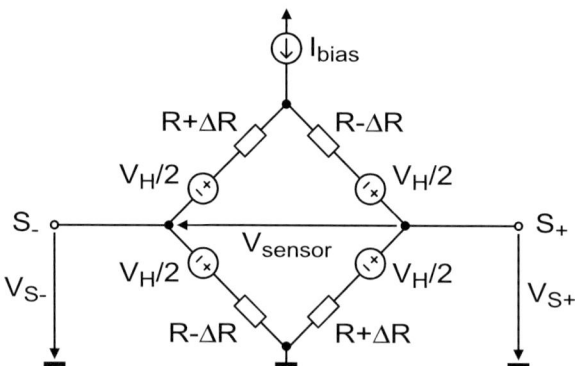

Figure 116. Modelling of the offset and Hall effect

The 4 additional voltage sources generate the Hall voltage according to equation 5.4 (or 5.6). The division by 2 accounts for the differential output mode. The sensor voltage V_{sensor} is:

Chapter 5: Hall microsystem with continuous calibration

$$V_{sensor} = V_H + V_{OS} = S_I I_{bias} B + \Delta R I_{bias} \qquad (5.9)$$

More elaborate models take into account the junction field effect, parasitic elements, etc. A complete reference can be found in [69].

3. SPINNING CURRENT TECHNIQUE

The spinning current technique allows to cancel the offset of Hall sensors [71][72]. The spinning current technique is the sensor counterpart of the chopper modulation in amplifiers (see chapter 2, section 3).

Since the sensor is totally symmetrical, its terminals can be exchanged. In particular, the current biasing and sensing role of the contact pairs C_+/C_- and S_+/S_-, respectively, can be swapped. This occurs if the sensor is alternately connected in the two configurations of figure 117.

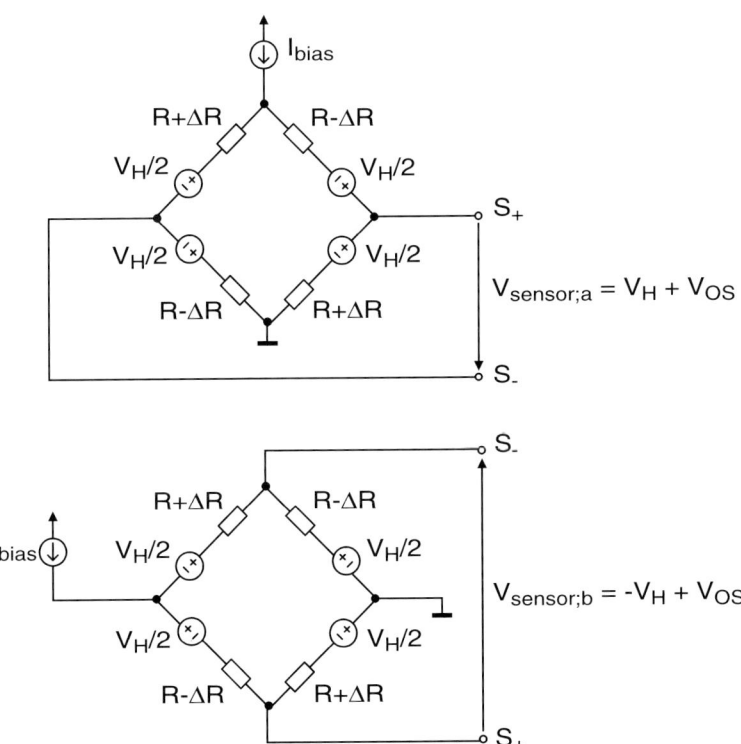

Figure 117. Spinning current technique
Top: Phase 1 (vertical biasing); *Bottom*: Phase 2 (horizontal biasing)

During the first phase, the sensor is biased vertically as in figure 116. The sensor voltage is the same as in equation 5.9:

$$V_{sensor;a} = V_H + V_{OS} \qquad (5.10)$$

During the second phase, the roles of the terminals are exchanged: The bias current flows horizontally and the voltage is sensed in the vertical direction. The sensor voltage is:

$$V_{sensor;b} = -V_H + V_{OS} \qquad (5.11)$$

If the phases are periodically alternated, the sensor produces exactly the same modulated voltage as at the output of the input modulator in the chopper amplifier presented in chapter 2, section 3: The sign of the signal voltage (the hall voltage V_H) is periodically reversed, whereas the sign of the offset V_{OS} remains constant. Using an appropriate demodulation (see chapter 2, section 3), the offset can be eliminated.

It is noteworthy that an amplifier used in conjunction with the sensor has its offset and 1/f noise also cancelled, exactly as in a chopper amplifier. This is because the offset and noise of the amplifier are added to V_{sensor} with a constant sign. Figure 118 presents a model of a sensor and differential preamplifier system. The terminals of the sensor are represented all on the same side for convenience.

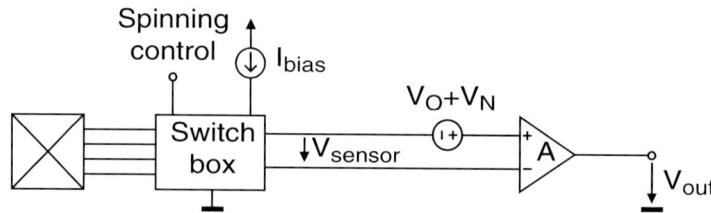

Figure 118. Sensor and preamplifier

Chapter 5: Hall microsystem with continuous calibration

The switch box implements the spinning current scheme of figure 117. It appropriately puts the sensor in one of the two configurations, based on the digital spinning control signal. Eight switches are necessary to allow the adequate connection of the 4 terminals of the sensor (2 switches per terminal) to the bias current source, the ground and the output connections in both configurations.

The system output voltage V_{out} is:

$$V_{out} = A(V_{sensor} + V_O + V_N) \tag{5.12}$$

During the first phase, this corresponds to:

$$V_{out;a} = A(V_{sensor;a} + V_O + V_N) = A(V_H + V_{OS} + V_O + V_N) \tag{5.13}$$

If the offset of the sensor is combined with the offset and noise of the amplifier, a resulting total input offset and noise $V_{off\&noise;in}$ can be calculated:

$$V_{off\&noise;in} = V_{OS} + V_O + V_N \tag{5.14}$$

Equation 5.13 can be rewritten:

$$V_{out;a} = AV_H + AV_{off\&noise;in} \tag{5.15}$$

During the second phase, the output voltage is:

$$V_{out;b} = A(V_{sensor;b} + V_O + V_N) = -AV_H + AV_{off\&noise;in} \tag{5.16}$$

The amplified offsets and the noise keep a constant sign during both phases, whereas the amplified Hall voltage is modulated. The total input offset and noise voltage, which is the combination of the offset of the sensor and the offset and noise of the amplifier, can thus be removed by demodulating the output voltage V_{out} as in conventional chopper amplifiers. In sampled data systems, this can be done by subtraction of the output values of both phases (equations 5.15 and 5.16), using for instance the circuit proposed in [73]. The result of the subtraction is:

$$V_{out} = V_{out;a} - V_{out;b} = 2AV_H \quad (5.17)$$

Both offsets of the sensor and of the amplifier are canceled. Furthermore, if the noise voltage remains constant between both phases, it is also removed. In practice, only the noise at lower frequencies than the spinning (modulation) frequency are removed, as explained in chapter 2, section 3.

4 SENSITIVITY CALIBRATION OF HALL SENSORS

The sensitivity of Hall sensors is not constant. If the Hall sensor is used in a closed-loop [74] measurement system, this is no issue since the feedback loop always brings the sensor back to its quiescent point, where neither non-linearity nor gain variations are problematic. In those systems, the sensor is used as a comparator and the measurement is derived from the amplitude of the feedback necessary to cancel the real physical quantity to be sensed.

In open-loop measurement systems, the sensitivity variations are problematic and are one of the major issues in the design of precise measurement systems based on Hall sensors.

Chapter 5: Hall microsystem with continuous calibration 161

4.1 Sensitivity drift of Hall sensors

The piezo-Hall effect [75] causes a modification of the current-related sensitivity when mechanical stresses are applied to the Hall sensor. The cause of these stresses are temperature variations, packaging and ageing. A complete reference can be found in [76].

Figure 119 shows the typical thermal drift of the current-related sensitivity S_I of an integrated Hall sensor [69] without packaging.

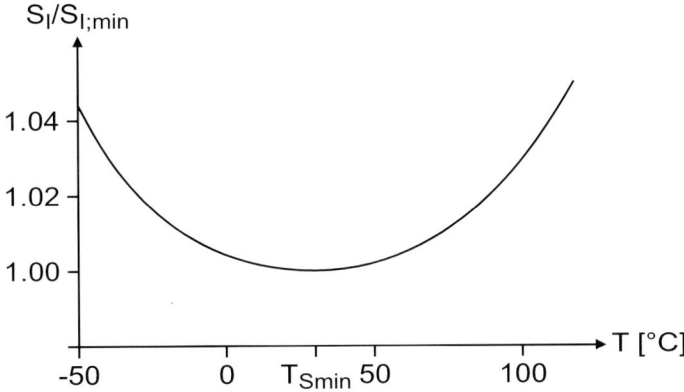

Figure 119. Typical thermal drift of the current-related sensitivity

The minimum gain value $S_{I;min}$ is reached for a temperature T_{Smin} usually comprised between 0 °C and 50 °C. Below and above T_{Smin}, the sensitivity increases by about 5 % for a temperature variation in the order of 100 °C. This corresponds to a variation in the order of 500 ppm/°C.

The drift is even higher if the sensor is encapsulated in a plastic package, because the difference between both thermal expansion coefficients of plastic and silicon causes additional thermo-mechanical stresses.

The temperature drift of Hall sensors can be compensated using a temperature sensor[1] and a calibration table containing the thermal characteristics of the sensitivity. It is also possible to use circuits or elements (resistors for instance) with opposite temperature coefficients to compensate the sensitivity drift of the sensor. Unfortunately, the history of mechanical stresses and temperature cycles causes an additional sensitivity modification of up to 2 % [75]. This drift is unpredictable and can thus not be compensated using a simple temperature compensation. The only solution to compensate the sensitivity drift is to measure it, compare it to a reference value, and apply an appropriate compensation.

[1]. The Hall sensor itself can be used as a temperature sensor, since its equivalent resistance seen between the current contacts changes as a function of temperature.

4.2 Integrated reference coils

To measure the current-related sensitivity of a Hall sensor, it must be placed in a magnetic field of known magnitude and biased with a known current. Using equation 5.4, the sensitivity can be calculated from the measured Hall voltage:

$$S_I = \frac{V_H}{I_{bias} B} \qquad (5.18)$$

A convenient manner to generate the reference magnetic field is to integrate a coil using one of the metal interconnection layers available in the fabrication process [77]. The idea of using the first metal layer to implement the coil is patented [78]. Figure 120 shows the top view of a Hall sensor and its overhanging calibration coil.

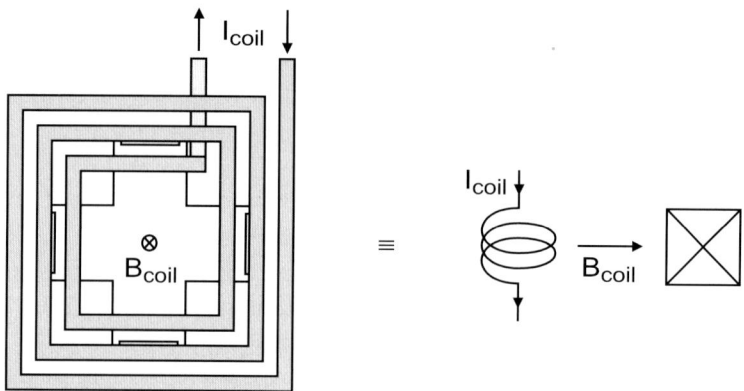

Figure 120.   Integrated calibration coil

The coil generates a vertical magnetic field which is measured by the sensor, along with an eventual external field (not represented). The equations for the calculation of the magnetic field generated by integrated coils can be found in [79] and [80].

To obtain a good efficiency, the coil should be located as closely as possible above the sensor. This parameter depends on the manufacturing process, but it can be influenced by choosing the first layer of interconnection metal (METAL1) if possible, since it is the closest to the sensor. On the other hand, it must also be considered that for high currents, the higher levels of metal (METAL2 for instance) are better suited [80], since they allow a higher current density.

Chapter 5: Hall microsystem with continuous calibration 163

A second important parameter is the number of turns and their arrangement. Since the efficiency of the inner turns is higher than the one of the outer turns, the number of turns should remain limited. In addition, the inner turns should be tight, i.e. close to the center of the sensor.

An important parameter of the coil is its current-related efficiency E_I, which is the ratio between the magnetic field and the coil current:

$$E_I = \frac{B_{coil}}{I_{coil}} \qquad (5.19)$$

In [81], an efficiency of 0.15 T/A is reported for a 3-turns coil. Higher values up to 0.4 T/A are achieved in [80]. The highest achievable magnetic field also depends on the maximum allowed current density in the coil, which limits the coil current to a few mA. As a result, the typical magnetic field generated by an integrated coil is in the order of 1 mT.

4.3   Sensitivity calibration

Using an integrated coil, it is possible to measure and calibrate the sensitivity of a Hall sensor. If the coil current I_{coil}, the sensor bias current I_{bias} and the coil sensitivity E_I are known, the sensor sensitivity S_I can be calculated by replacing equation 5.19 in 5.18:

$$S_I = \frac{V_H}{I_{bias} B_{coil}} = \frac{V_H}{I_{bias} E_I I_{coil}} \qquad (5.20)$$

Several calibration techniques using an integrated coil already exist. They differ by the moment where the calibration is performed, and the circuit topology used for the calibration.

The simplest calibration technique consists in performing a single calibration after production to adjust the gain of the circuit for its entire lifetime. This technique allows to adjust the nominal gain at the calibration temperature (usually room temperature), but neither compensates temperature effects nor ageing. If a first-order temperature compensation is further implemented, the sensitivity drift due to temperature can be reduced from 500 ppm/°C to about 300 ppm/°C [82]. This limit is due to second-order temperature effects, which can be canceled using a calibration table instead of a first-order compensation. This implies, however, that the sensitivity of the sensor must be measured at different temperatures, which is almost impracticable in production environments. Furthermore, this technique does not compensate effects due to ageing and purely mechanical constraints, which can affect the sensitivity up to 2 % [75].

To overcome these drawbacks, the sensor can be calibrated at regular intervals. If the sensitivity of the sensor is measured in its real operating environment, it can be accurately compensated, taking into account any cause of the drift (temperature, mechanical constraint, ageing, etc.). Using this technique, the temperature drift is reduced to 100 ppm/°C [83].

Figure 121 presents a basic system performing sensitivity calibration.

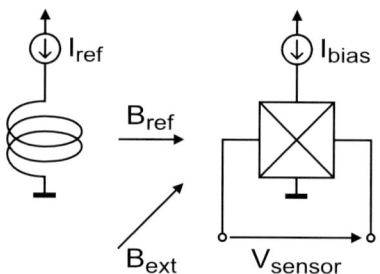

Figure 121. Sensitivity calibration principle

The Hall sensor measures the external magnetic field B_{ext}, to which a reference magnetic field B_{ref} is added for calibration. The sensor output voltage V_{sensor} is calculated from equation 5.9:

$$V_{sensor} = V_H + V_{OS} = S_I I_{bias}(B_{ext} + B_{ref}) + V_{OS} \quad (5.21)$$

If two different reference magnetic fields $B_{ref;1}$ and $B_{ref;2}$ are successively applied, and the corresponding sensor voltages $V_{sensor;1}$ and $V_{sensor;2}$ measured, the sensitivity S_I of the Hall sensor can be calculated. The subtraction of the two voltage measurements indeed gives:

$$\begin{aligned} V_{sensor;1} - V_{sensor;2} &= S_I I_{bias}(B_{ext} + B_{ref;1}) + V_{OS} \\ &\quad -[S_I I_{bias}(B_{ext} + B_{ref;2}) + V_{OS}] \\ &= S_I I_{bias}(B_{ref;1} - B_{ref;2}) \end{aligned} \quad (5.22)$$

The offset and external magnetic field cancel, and the sensitivity can be extracted:

$$S_I = \frac{V_{sensor;1} - V_{sensor;2}}{I_{bias}(B_{ref;1} - B_{ref;2})} \quad (5.23)$$

Chapter 5: Hall microsystem with continuous calibration

Using equation 5.19 to replace in equation 5.23 both successive reference magnetic fields $B_{ref;1}$ and $B_{ref;2}$ by their corresponding coil currents $I_{ref;1}$ and $I_{ref;2}$, the sensitivity finally is:

$$S_I = \frac{V_{sensor;1} - V_{sensor;2}}{I_{bias} E_I (I_{ref;1} - I_{ref;2})} \quad (5.24)$$

This equation is similar to equation 5.20, but it has the advantage of cancelling the effect of both the offset of the sensor and the external magnetic field. However, the latter has no effect in equation 5.24 *only if it remains constant* during the calibration, i.e. its value is equal for both measurements. If this assumption is not true, a parasitic term due to the variation of the external field ΔB_{ext} distorts the sensitivity calculation:

$$S_I = \frac{V_{sensor;1} - V_{sensor;2}}{I_{bias}[E_I (I_{ref;1} - I_{ref;2}) + \Delta B_{ext}]} \quad (5.25)$$

where ΔB_{ext} is calculated as the difference between the external magnetic fields $B_{ext;1}$ and $B_{ext;2}$ at the instant of the first and second measurements respectively:

$$\Delta B_{ext} = B_{ext;1} - B_{ext;2} \quad (5.26)$$

The amplitude of the term ΔB_{ext} depends on the highest frequency component (or bandwidth) of B_{ext} with respect to the time interval between the two measurements. Ideally, the calibration period should be much smaller than the period of B_{ext} (of its highest frequency component). Figure 122 shows a graphical representation of a periodical signal and emphasizes the influence of the calibration period on the term ΔB_{ext}.

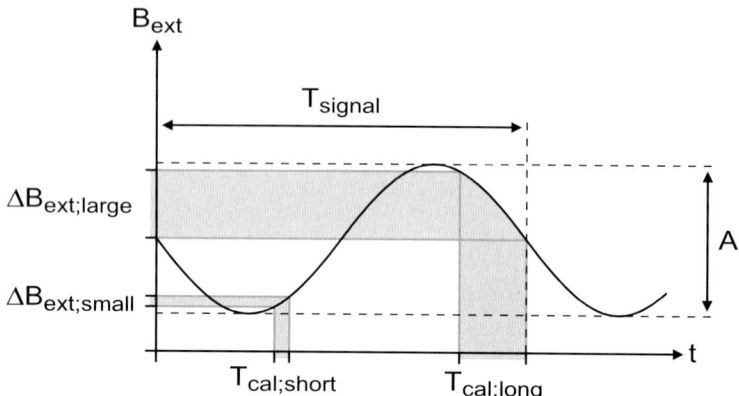

Figure 122. Influence of the calibration period on the variation of B_{ext}

If a long calibration period $T_{cal;long}$ with respect to the signal period T_{signal} is used, the variation of B_{ext} between the beginning and end of the calibration (ΔB_{ext}) is also large ($\Delta B_{ext;large}$). If a shorter period $T_{cal;short}$ is chosen, the variation is smaller ($\Delta B_{ext;small}$).

The second parameter that influences ΔB_{ext} is obviously the amplitude A of the global variation of B_{ext}, since it multiplies the relative variation (see figure 122).

In fact, the parasitic term ΔB_{ext} in equation 5.25 has to be related to the amplitude of the reference magnetic field B_{ref}. The calculation of the sensitivity is accurate if ΔB_{ext} remains small with respect to the term $B_{ref;1} - B_{ref;2}$. Whereas the calibration period T_{cal} can usually be decreased in order to minimize ΔB_{ext}, the amplitude ratio of B_{ref} to the absolute amplitude A of B_{ext} is influenced with difficulty. As shown in section 4.2, an integrated coil can generate reference magnetic fields up to 1 mT, whereas the amplitude of the external magnetic field in applications of Hall sensors can be up to 100 mT. The ratio between these amplitudes is unfavorable but still allows the development of calibration systems based on this principle.

4.4 State of the art

Different circuit topologies implementing the calibration principle presented in section 4.3 have already been implemented. In order to produce a continuous output signal and perform calibration at the same time, i.e. to continuously generate a user signal (corresponding to B_{ext}) and a reference signal (corresponding to B_{ref}), they use two *matched sensors*. As for other integrated components, the careful layout and the rigorous respect of matching rules

Chapter 5: Hall microsystem with continuous calibration

guarantees that the characteristics of both sensors, in particular the sensitivity, are similar.

Two categories of circuits can be identified when two sensors are used. Either both calibration and external fields are measured by both sensors (in two different combinations), and their signals processed to extract the calibration and user signal, or each sensor is used to process separately either the calibration or external field.

Figure 123 presents the solution where the signals are combined for both sensors. This is one of the circuit topologies presented in [84].

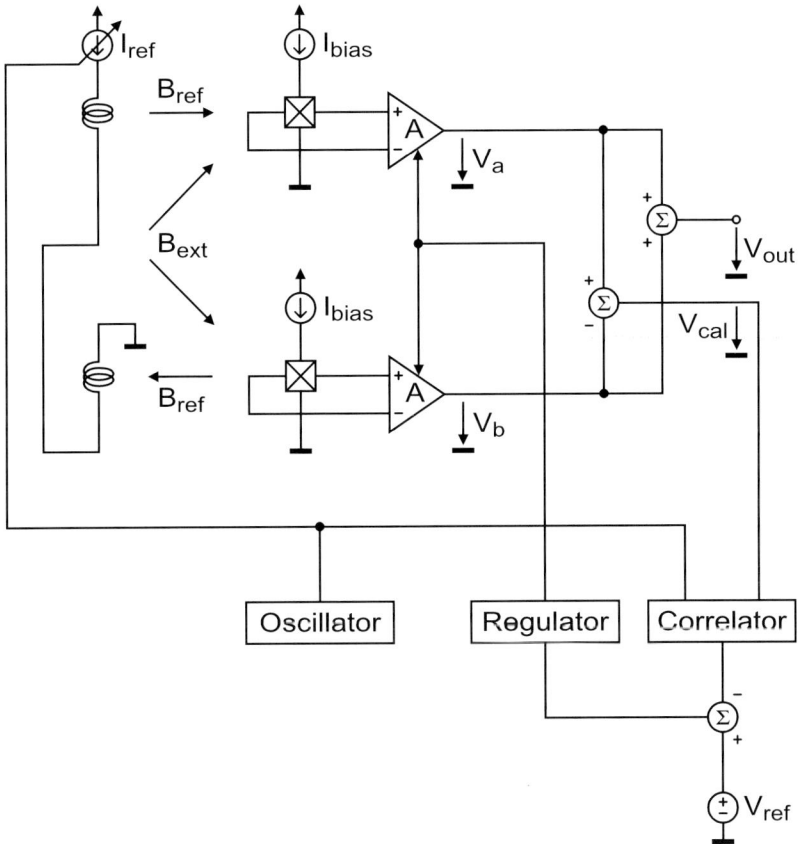

Figure 123. Calibration by dual signal ± reference measurement paths

The two parallel measurement paths measure dual signals. Both sensors are submitted to the external magnetic field B_{ext}, but also to the reference magnetic field B_{ref}. The latter however is measured with an opposite sign in both channels. The voltage V_a at the output of the amplifier in the first path is:

$$V_a = AS_I I_{bias}(B_{ext} + B_{ref}) \qquad (5.27)$$

At the output of the second channel, the voltage V_b is:

$$V_b = AS_I I_{bias}(B_{ext} - B_{ref}) \qquad (5.28)$$

If the results of both paths are added, the user output signal V_{out} is obtained:

$$V_{out} = V_a + V_b = 2AS_I I_{bias} B_{ext} \qquad (5.29)$$

The reference field is cancelled by the addition, and V_{out} contains only the amplified external field component. If, on the other hand, the results of both channels are subtracted, the calibration signal V_{cal} is obtained:

$$V_{cal} = V_a - V_b = 2AS_I I_{bias} B_{ref} \qquad (5.30)$$

If the reference magnetic field B_{ref} is modulated by an oscillator, it is possible to extract a continuous calibration signal by using a correlator, which can then be compared to a reference voltage V_{ref}. A regulator adjusts the gain A of the amplifiers to maintain the difference null. V_{ref} is thus the control signal setting the nominal gain of the sensors. It is also possible to adjust the gain of the system by modifying the sensor bias current I_{bias}, since both I_{bias} and A have the same multiplicative effect in equations 5.27 and 5.28. In fact, the compensation feedback loop adjusts the overall gain of the sensor and its preamplifier.

The advantage of modulating the reference signal is that the reference can be placed at frequency outside the user bandwidth (the frequency range of B_{ext}). If there is a parasitic component due to B_{ext} in the subtraction (because both channels are not perfectly matched), it can be filtered out. Furthermore, if the spinning current is not implemented, the reference modulation allows to cancel the offset and 1/f noise in the reference signal after the correlator.

Figure 124 presents another circuit topology [85], where the measurement paths for the external and reference signals are separated.

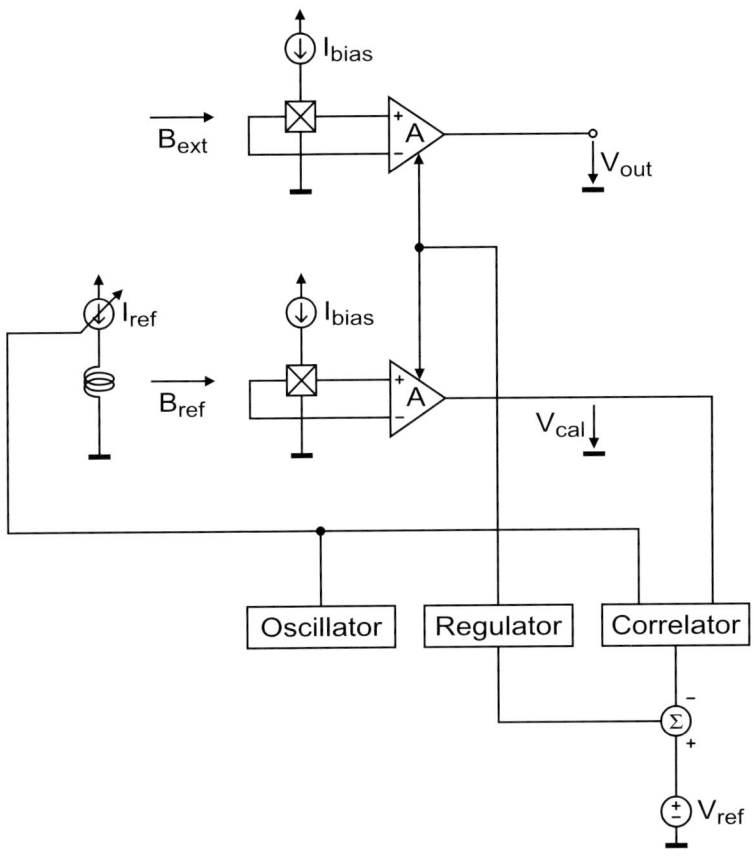

Figure 124. Calibration by separate signal and reference measurement paths

In the upper path, no calibration coil is integrated above the Hall sensor, which thus senses only the external magnetic field:

$$V_{out} = AS_I I_{bias} B_{ext} \qquad (5.31)$$

Using an array of Hall sensors instead of a single one (see chapter 6, section 2) and a special combination of coils [80], it is possible to remove the influence of the external magnetic field B_{ext} in the second calibration path. Using two Hall sensors for instance, this is achieved by applying opposite calibration fields to both sensors. Then, the subtraction of both sensor voltages eliminates B_{ext}, in a similar way as in equation 5.30. The calibration voltage V_{cal} is thus:

$$V_{cal} = AS_I I_{bias} B_{ref} \qquad (5.32)$$

For the same reason as in the dual channel circuit of figure 123, the reference magnetic field is modulated. The modulation frequency is 1 kHz in [80], because of the limitation due to the capacitive parasitic coupling between the calibration coil and the sensor.

Another element integrated in the feedback loop in [80] is a low-pass filter with a very low cutoff frequency (0.5 Hz). By reducing the bandwidth of the calibration signal after the correlator to a small range, the filter increases the signal-to-noise ratio (SNR). This is necessary because of the small signal level generated by the Hall sensor in response to the weak calibration field. After the correlator, the calibration signal is DC and is not attenuated by the low-pass filter.

A topology using a single sensor to measure both B_{ext} and B_{ref} is presented in [86]. In this circuit, B_{ref} is modulated at much higher frequency than the bandwidth of B_{ext} and the signals are separated in the spectral domain. It is then possible to extract both external and reference signals from a single channel by low- and high-pass filtering respectively. Figure 125 presents the corresponding circuit.

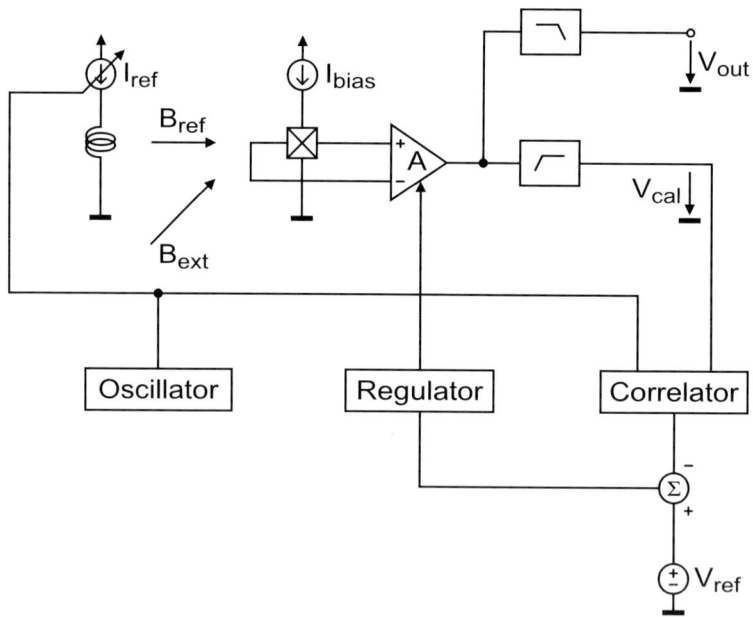

Figure 125. Calibration by frequency separation

The disadvantage of this topology is that it cannot be used when the bandwidth of B_{ext} is large, because it implies either a very high reference frequency, or very selective filters to accurately separate both components.

To measure the sensitivity of Hall sensors, the techniques presented in this section use spatial separation or frequency separation of the external and reference signals. The first category requires the use of a matched pair of sensors and amplifiers, whereas the second one is not applicable to high bandwidth systems.

At the end of this chapter, a new topology combining the advantages of the techniques discussed in this section is presented. By using *time multiplexing*, it allows a single sensor system based on the spinning current technique to continuously measure and calibrate the sensitivity of the sensor, even in applications requiring a high bandwidth.

5 HALL SENSOR MICROSYSTEMS

There are various applications for microsystems combining an integrated Hall sensor and its analog front end [62]. They are used in position sensors [87], angular encoders [52] and general high-resolution magnetic sensors [88][89]. They also allow the shuntless measurement of electrical currents [90], and by extension of electrical power. In fact, the Ampere-Laplace law states that an electrical current generates a magnetic field. By the measurement of the magnetic field near a conductor, the current flowing through it can be deduced. If furthermore the voltage in the conductor is probed, the electrical power can be calculated as the current multiplied by the voltage.

Because they allow the contactless and shuntless measurement of the current, Hall sensor microsystems are well-suited for high-current applications. It indeed avoids to interrupt the circuit under measurement, and above all to insert an energy dissipative shunt in high current power lines.

5.1 Analog front-ends for current measurement

The typical specifications of a Hall sensor microsystem for current measurement [82] are summarized in table 14.

Table 14. Typical specifications of a current measurement microsystem

Parameter	Value			Unit
	Min.	Typ.	Max.	
Temperature range	-40	-	125	°C
Supply voltage	4.75	5	5.25	V
Power consumption	-	100	-	mW
Magnetic field measurement range	-50	-	50	mT
Precision	-1	-	1	%
Bandwidth	0	-	30	kHz
Response time	-	-	3	µs
Input-referred offset	-10	-	10	µT
Input-referred RMS noise (BW = 1 MHz)	-	10	-	mV
Gain	-	40	-	V/T
Temperature drift of the gain	-	300	-	ppm/°C

Concerning the operating conditions, the main constraint is the temperature range. Since one major application of the current measurement ASICs (Application Specific Integrated Circuit) is the automotive industry (battery monitoring, electric motor control, etc.), they can be submitted to high temperature variations. For other applications, the specification can be relaxed.

The bandwidth of the system is not very extended. However, the response time of the circuit must be short for some applications where short-circuits have to be detected.

The offset and noise are greatly reduced by the spinning current technique (section 3). On the other hand, the temperature drift of the system gain is limited by the sensitivity drift of the integrated Hall sensor. Some of the techniques presented in section 4.4 can be applied to the current measurement application by using two parallel sensor systems and combining their results. In the next section, a new calibration technique using a single sensor is presented.

Chapter 5: Hall microsystem with continuous calibration 173

6 CONTINUOUS DIGITAL GAIN CALIBRATION TECHNIQUE

The technique presented in this section allows the continuous calibration of the gain of a Hall sensor analog front end for current measurement [91][92]. The gain is determined using an integrated coil which generates an appropriately modulated reference magnetic field. The reference field is measured by the sensor and amplified jointly with the external magnetic field. Both signals are extracted after the preamplifier by two separate demodulators. The demodulated calibration signal is compared to a reference value and the gain of the sensor and preamplifier is appropriately compensated.

Since the gain measurement is performed directly and continuously on the Hall sensor measuring the user magnetic field, the compensation cancels any sensitivity drift, even those due to purely mechanical stresses, second-order temperature effects and ageing.

This chapter presents the system architecture and its functioning. The implementation is presented in chapter 6.

6.1 Principle

The calibration system is based on the extension of the spinning current technique. The idea is to combine the modulation of the reference signal with the spinning current modulation, by using two modulation frequencies that are integer multiples/divisors of another. The modulated signal successively contains 4 different combinations of the external field and the reference field, which can be extracted by appropriate synchronous demodulation. Figure 126 presents the general system architecture.

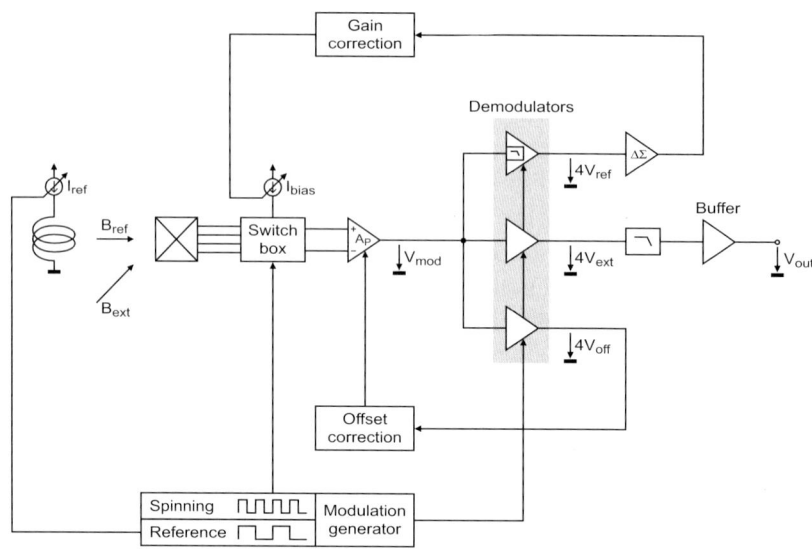

Figure 126. System architecture

The clock generator produces both modulation signals for the spinning current implemented by the switch box and the reference coil. In this example, the reference is modulated at half the spinning frequency, but other modulation schemes can be implemented (chapter 6, section 6.4).

Three parallel demodulators extract three different signals from the preamplified modulated signal V_{mod}: the user signal voltage V_{ext} (corresponding to the external magnetic field), the gain calibration signal V_{ref} and the offset signal V_{off}. This is done by using a different demodulation scheme for each channel. The three signals are multiplied by a constant factor 4 due to the demodulation principle (section 6.3).

The user signal is low-pass filtered and buffered to respectively eliminate the high-frequency component due to the modulation (as in chopper systems presented in chapter 2, section 3) and provide sufficient output current. The voltage V_{out} is the output of the current measurement system.

The gain and offset informations are used in feedback correction loops to compensate the imperfections of the sensor and the preamplifier. A low-pass filter is included in the reference demodulator to increase the signal-to-noise ratio. The filtering is second-order because the demodulator is followed by a second low-pass filter implemented by a delta-sigma ($\Delta\Sigma$) converter, which also converts the calibration signal to the digital domain.

6.2 Combined modulation scheme

As stated in section 3, the spinning current technique produces a periodic sign change of the preamplifier output voltage, whereas the offset remains constant. In the circuit of figure 126, the output V_{mod} of the preamplifier is (using equations 5.4, 5.15 and 5.16):

$$V_{mod} = \pm A_P S_I I_{bias} B + A_P V_{off;in} \quad (5.33)$$

where A_P is gain of the preamplifier, S_I and I_{bias} the current-related sensitivity and bias current of the Hall sensor, respectively, and $V_{off;in}$ is the total input offset, comprising both the offsets of the Hall sensor (V_{OS}) and of the preamplifier (V_O):

$$V_{off;in} = V_{OS} + V_O \quad (5.34)$$

The noise term V_N of equation 5.14 is omitted to facilitate understanding.

Exactly as the spinning current reverses the signal component of the sensor, the modulation of the reference current I_{ref} periodically changes the direction of the current flowing through the coil. Using switches, the coil is reversed alternately to generate a magnetic field B_{ref} of constant amplitude, but opposite direction. The total magnetic field B measured by the sensor is thus:

$$B = B_{ext} \pm B_{ref} \quad (5.35)$$

Replacing equation 5.35 in 5.33 gives:

$$V_{mod} = \pm A_P S_I I_{bias} (B_{ext} \pm B_{ref}) + A_P V_{off;in} \quad (5.36)$$

To simplify the notation, the voltage equivalent signal of each component at the output of the preamplifier is introduced. The external component V_{ext} is:

$$V_{ext} = A_P S_I I_{bias} B_{ext} \quad (5.37)$$

The reference component V_{ref} is:

$$V_{ref} = A_P S_I I_{bias} B_{ref} \quad (5.38)$$

Finally, the amplified voltage V_{off} is:

$$V_{off} = A_p V_{off;in} \qquad (5.39)$$

Replacing these 3 voltages in equation 5.36 gives:

$$V_{mod} = \pm(V_{ext} \pm V_{ref}) + V_{off} \qquad (5.40)$$

The two ± operators represent the effect of both modulations. The first one (outside the parentheses) corresponds to the spinning current and indeed influences both external and reference components. The internal ± operator accounts for the direction of the reference magnetic field which is periodically reversed and is consequently either added to or subtracted from the external component.

Since both modulations are performed according to the schematic representation of figure 126, i.e. with a reference modulation at half the spinning current frequency, there are 4 successive modulation phases. Each of these phases corresponds to one of the 4 possible combinations of ± signs in equation 5.40. Table 15 presents the voltage output value V_{mod} of the preamplifier for each phase.

Table 15. Combined modulation scheme

Phase (i)	Modulation		Preamplifier output ($V_{mod;i}$)
	Spinning	Reference	
1	+	+	$+ (V_{ext} + V_{ref}) + V_{off}$
2	−	+	$- (V_{ext} + V_{ref}) + V_{off}$
3	+	−	$+ (V_{ext} - V_{ref}) + V_{off}$
4	−	−	$- (V_{ext} - V_{ref}) + V_{off}$

The notation $V_{mod;i}$ is used to represent the output of the preamplifier during phase number i.

6.3 Demodulation schemes

If the values $V_{mod;i}$ of each phase of the modulation are combined using additions and subtractions, the three signal components can be extracted. The component obtained depends on the sign sequence used for the demodulation.

Chapter 5: Hall microsystem with continuous calibration

The component corresponding to the external magnetic field is calculated as:

$$V_{mod;1} - V_{mod;2} + V_{mod;3} - V_{mod;4} = 4V_{ext} \quad (5.41)$$

To obtain the reference signal, the signs are reversed for phases 3 and 4:

$$V_{mod;1} - V_{mod;2} - V_{mod;3} + V_{mod;4} = 4V_{ref} \quad (5.42)$$

Finally, the offset voltage is the mean value (DC component) of V_{mod} and is thus extracted by summing the 4 phases:

$$V_{mod;1} + V_{mod;2} + V_{mod;3} + V_{mod;4} = 4V_{ref} \quad (5.43)$$

Table 16 summarizes the three different demodulation schemes. The preamplifier modulated output values $V_{mod;i}$ are repeated for convenience.

Table 16. Demodulation schemes

Phase (i)	Preamplifier output ($V_{mod;i}$)	Demodulation		
		Signal	Reference	Offset
1	$+ (V_{ext} + V_{ref}) + V_{off}$	+	+	+
2	$- (V_{ext} + V_{ref}) + V_{off}$	-	-	+
3	$+ (V_{ext} - V_{ref}) + V_{off}$	+	-	+
4	$- (V_{ext} - V_{ref}) + V_{off}$	-	+	+
		$4 V_{ext}$	$4 V_{ref}$	$4 V_{off}$

If 3 demodulators are connected in parallel to the output of the preamplifier, each one applying one of the demodulation schemes corresponding to the columns of table 16, the 3 signal components V_{ext}, V_{ref} and V_{off} are available continuously and simultaneously. A new value is calculated for each of them during 4 consecutive phases, and the values are updated every 4th phase.

If some of the 3 signals must not be available simultaneously, a single demodulator can be used to implement multiple demodulation schemes alternately, generating successively different signals. This option is interesting for the offset and gain (reference) signal components, since their respective feedback loops can bear a lower signal rate and/or periodic interruption.

If the offset signal does not have to be extracted, a scheme producing intermediate results can also be used. It allows to share a part of the circuit of two parallel demodulators. In fact, the partial subtraction result $V_{mod;1/2}$ of phases 1 and 2 combined (corresponding to the half-period of the reference modulation) is:

$$V_{mod;1/2} = V_{mod;1} - V_{mod;2} = 2(V_{ext} + V_{ref}) \qquad (5.44)$$

The second partial subtraction $V_{mod;3/4}$ of phases 3 and 4 gives:

$$V_{mod;3/4} = V_{mod;3} - V_{mod;4} = 2(V_{ext} - V_{ref}) \qquad (5.45)$$

In both partial results, the offset component is cancelled. They can then be combined by addition or subtraction. The component corresponding to the external magnetic field is extracted by addition:

$$V_{mod;1/2} + V_{mod;3/4} = 4V_{ext} \qquad (5.46)$$

If on the other hand a subtraction of the partial results is performed, the reference signal is obtained:

$$V_{mod;1/2} - V_{mod;3/4} = 4V_{ref} \qquad (5.47)$$

Another advantageous result is obtained using a similar technique and a shifted combination of phases 2/3 and 1/4. It allows the extraction of the external signal components from V_{mod} every 2 phases instead of 4 as presented above. In fact, during phases 2 and 3:

$$-V_{mod;2} + V_{mod;3} = 2V_{ext} \qquad (5.48)$$

The same result is obtained for the other pair 1/4 of successive phases (4/1, actually, since phase 1 in a new cycle comes just after phase 4 of the preceding one):

$$-V_{mod;4} + V_{mod;1} = 2V_{ext} \qquad (5.49)$$

If this technique is used for the external component, it allows the doubling of the user signal output rate, and thus also a doubled bandwidth and a halved response time, which are important characteristics of current measurement microsystems.

There are still other possible modulation and demodulations schemes, which can be used to improve other system characteristics. They are presented in chapter 6, section 6.4.

6.4 Gain compensation

The goal of the gain compensation system is to maintain a stable value of the sensitivity of the overall magnetic field measurement circuit, i.e. viewed from the user output signal V_{out}, or equivalently from its unfiltered alias $4V_{ext}$. The relation between V_{out} (neglecting the effect of the low-pass filtering) and B_{ext} is (using equation 5.37):

$$V_{out} = 4V_{ext} = 4A_p S_I I_{bias} B_{ext} = KB_{ext} \qquad (5.50)$$

The relation is linear, and the proportionality factor is the global system sensitivity K [V/T]:

$$K = 4A_p S_I I_{bias} \qquad (5.51)$$

The sensitivity, and consequently its drift, linearly depends on the preamplifier gain and the current-related sensitivity and bias current of the Hall sensor. As explained in section 4.1, the current-related sensitivity drift due to temperature variations, mechanical stresses and ageing is problematic. Equation 5.51 shows that in order to keep the global sensitivity K constant, a drift of S_I must be compensated by an opposite variation of the gain of the preamplifier A_p, the sensor bias current I_{bias}, or a combination of both. In the present microsystem, the correction is done by adjusting I_{bias}.

If the sensitivity K must be compensated by the gain calibration in order to remain constant, this implies that it is measured and compared to a stable reference.

The measurement of the sensitivity is performed through the reference signal V_{ref}. At the output of the reference demodulator, the signal is (using equations 5.19 and 5.38):

$$4V_{ref} = 4A_p S_I I_{bias} B_{ref} = 4A_p S_I I_{bias} E_I I_{ref} = KE_I I_{ref} \qquad (5.52)$$

If the current-related efficiency E_I and the bias current I_{ref} of the coil are constant, keeping K stable is equivalent to keeping the output value of the reference demodulator ($4V_{ref}$) constant. Concerning E_I, this stability assumption is correct [80]. For I_{ref}, a careful design must guarantee its stability or its link to the nominal value V_{nom} to which the output of the reference demodulator is compared in the adjustment loop [86].

Figure 127 shows the structure of the gain adjustment feedback loop.

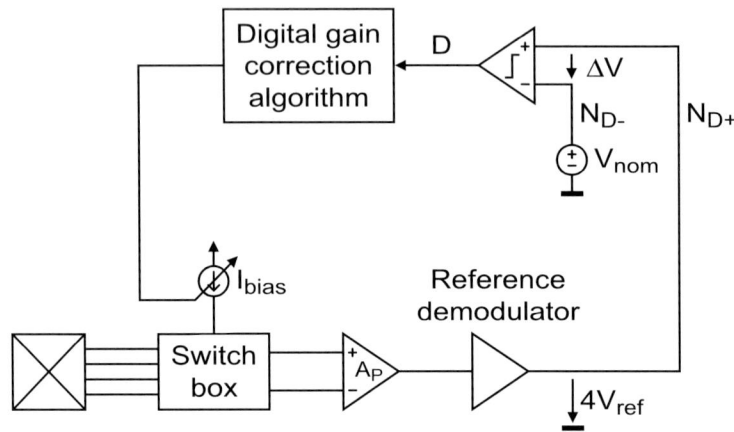

Figure 127. Gain adjustment feedback loop

The output of the reference demodulator ($4V_{ref}$) is compared to the nominal reference value V_{nom}. If the sensitivity K has a nominal value K_{nom}, to which is added a drift ΔK:

$$K = K_{nom} + \Delta K \tag{5.53}$$

If furthermore V_{nom} is chosen so that:

$$V_{nom} = K_{nom} E_I I_{ref} \tag{5.54}$$

The voltage difference ΔV is in this case:

$$\Delta V = 4V_{ref} - V_{nom} = (K_{nom} + \Delta K) E_I I_{ref} - K_{nom} E_I I_{ref} = \Delta K E_I I_{ref} \tag{5.55}$$

If the compensation feedback loop maintains $\Delta V = 0$, this implies that:

$$\Delta K = \frac{\Delta V}{E_I I_{ref}} = 0 \tag{5.56}$$

and thus that $K = K_{nom}$. In other words, the pair N_{D+} and N_{D-} are gain drift *detection nodes* in the sense of chapter 4, section 3.2, ΔV ($= \delta$) is the *detec-*

Chapter 5: Hall microsystem with continuous calibration

tion signal and the digital output signal D of the comparator can be used as *decision signal* by a digital compensation circuit to compensate the offset.

In the circuit of figure 127, the comparison of the measured value to the nominal value is performed in the analog domain. In the current measurement microsystem presented in figure 126, the comparison is equivalently done digitally. Figure 128 shows a digital implementation of the feedback loop.

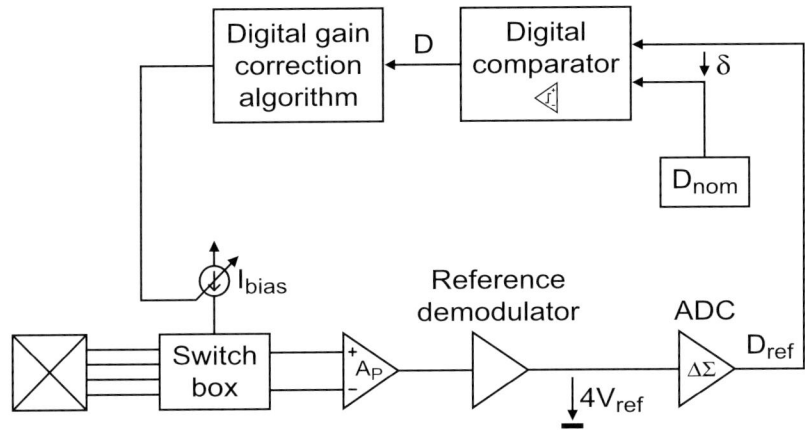

Figure 128. Gain adjustment feedback loop with ADC and digital comparison

The $4V_{ref}$ value is converted by the delta-sigma analog-to-digital converter (ADC) into the digital word D_{ref}. The latter is compared digitally to the nominal value D_{nom} corresponding to its analog counterpart V_{nom}. The detection signal δ is in this case the difference between the digital values. Both the analog and the digital implementations are equivalent. However, in the circuit of figure 126, the digital version is more advantageous because it allows the indefinite storage of the nominal value (see also chapter 2, sections 4.2 and 8).

Concerning the means of the compensation, two choices are possible: adjusting I_{bias} or A_P. Since the gain of the preamplifier is fixed by the resistor ratio in the feedback path of operational amplifiers, adjusting A_P precisely is difficult. It can be done by using a resistor array (see chapter 3, section 5). On the other hand, the adjustment of I_{bias} is more advantageous. Since the compensation value is a current, it can be implemented using a sub-binary current-mode DAC like a $M/2^+M$ (chapter 3, section 11) for instance. The DAC injects a compensation current in the *compensation node* (see also chapter 4, section 3.3), which is in this case the sensor bias input.

The compensation DAC can be directly used to generate the total sensor bias current I_{bias}, comprising both the nominal bias current I_{nom} of the sensor and the compensation current I_{comp}. However, it is more advantageous to implement I_{nom} separately, as shown in figure 129. The constant current source I_{nom} generates the main part of the bias current (corresponding to K_{nom} in equation 5.53), whereas the adjustable compensation current I_{comp} corresponds only to the drift ΔK. The total bias current of the sensor is:

$$I_{bias} = I_{nom} + I_{comp} \qquad (5.57)$$

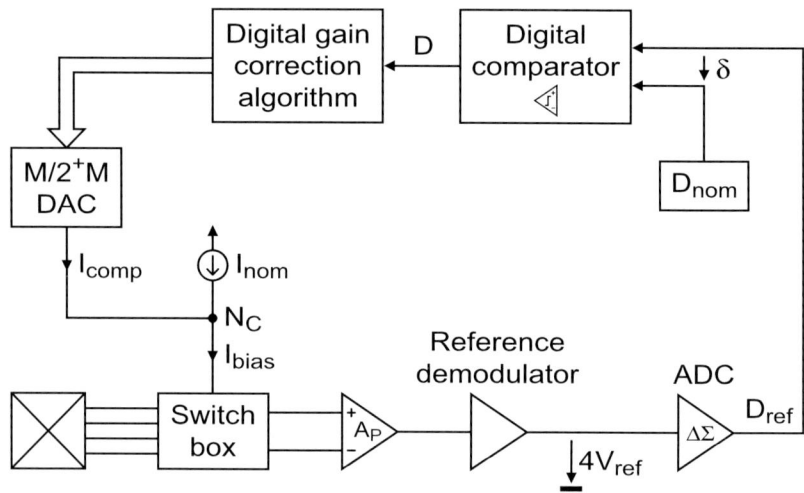

Figure 129. Compensation current injection

In figure 129, the *compensation node* N_C (see also chapter 4, section 3.3), where the compensation current I_{comp} is injected, is the sensor bias current input (through the switch box realizing the spinning current).

By implementing the nominal bias current I_{nom} separately, the full scale of the DAC and consequently also its necessary resolution are reduced, in accordance with the analysis of chapter 4, section 3.4.

The digital gain correction circuit uses the decision signal D to appropriately adjust the digital input word of the compensation DAC. In this application, the signal output is sampled at high frequency and does not allow intermediate calibration phases. It is also not acceptable to generate large temporary sensitivity variations (K) by using directly a successive approximations algorithm (see chapter 4, section 3.6). For these reasons, the digital-to-analog converter generating I_{comp} and the associated algorithm must be the up/down circuit presented in chapter 4, section 3.7 or an equivalent system.

6.5 Offset compensation

The typical Hall sensor voltage level is low, in the order of 4 mV for a full measurement scale of 50 mT. This value is comparable to the input-referred offset of the sensor and preamplifier. Furthermore, the gain of the preamplifier is chosen to obtain a high signal level V_{mod} for the demodulators. Consequently, the offset can saturate the output of the preamplifier if no offset correction is performed.

The offset compensation is performed in the same manner as the gain compensation presented in section 6.4. Figure 130 presents the offset correction feedback loop, which is similar to the topologies proposed in [93] and [94].

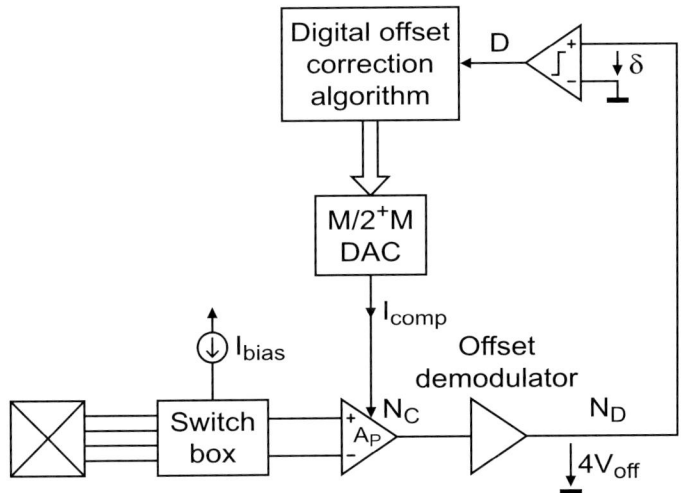

Figure 130. Offset correction feedback loop

The detection node N_D is the output of the offset demodulator. The detection signal δ is directly the voltage in this node with respect to ground. It is calculated using equations 5.34 and 5.39:

$$\delta = 4V_{off} = 4A_P V_{off;in} = 4A_P(V_{OS} + V_O) \qquad (5.58)$$

The digital offset correction algorithm increases or decreases the compensation current I_{comp} which is injected into an internal offset compensation node N_C of the preamplifier (see chapter 4, section 3.3), basing its decision on the digital output D of the comparator. If the compensation is perfect, $\delta = 0$ and the total input-referred offset $V_{off;in}$, which is the sum of the sensor offset voltage V_{OS} and preamplifier input-referred offset V_O is also null (from equation 5.58).

6.6 Noise filtering

Another problem arising from the low Hall sensor voltage level is the signal-to-noise ratio, in particular for the reference field measurement. In fact, the reasonably achievable magnetic field with an integrated coil is 0.5 mT. This is 100 times smaller than the typical full measurement scale of 50 mT. In order to adjust the gain with a good precision, a fraction of the reference signal must be detectable in the modulated signal V_{mod} at the output of the preamplifier. To be coherent, this fraction should be much smaller than the precision of the system, which is typically 1 % (see table 14). A gain regulation with a 0.1 % precision is thus reasonable. But this 0.1 % fraction of the reference field $V_{ref;0.1\%}$ corresponds to a magnetic field of only 0.5 µT.

Table 17 summarizes the signal levels corresponding to the external magnetic field, the reference field, and its 0.1 % fraction. The levels are expressed as magnetic fields (T unit) and corresponding Hall voltage between the sensor terminals. A sensor sensitivity of 0.08 V/T is assumed, which is typical for integrated Hall sensors and corresponds to a current-related sensitivity $S_I = 80$ V/TA and a bias current I_{bias} of 1 mA.

Table 17. External signal, reference signal and noise levels

Signal	Magnetic field	Unit	Hall voltage	Unit
External field	50	mT	4	mV
Reference field	0.5	mT	40	µV
0.1 % of reference field	0.5	µT	40	nV
White noise	-	-	20	nV√Hz

Another information present in table 17 is the white noise floor level $V_{n;white}$ (the root spectral density) of the sensor and preamplifier, referred to the input of the preamplifier. This value is important with respect to the signal level. In fact, in order to allow the extraction of the 0.1 % fraction of the reference signal from V_{mod}, it is obviously mandatory that its level be above the noise level. Since the combined modulation is performed at high frequency, the noise level at the modulation frequency is dominated by the white noise as in chopper systems (see chapter 2, section 3). Seen from the input of the preamplifier, this signifies that the 0.1 % fraction of the reference field signal

$V_{ref;in;0.1\%}$ must be higher than the white noise floor level $V_{n;white}$. A reasonably achievable value for the white noise floor is 20 nV√Hz [95][89].

Figure 131 shows a spectral representation of the input-referred 0.1 % fraction of the modulated reference signal, combined with the noise, using the values from table 17.

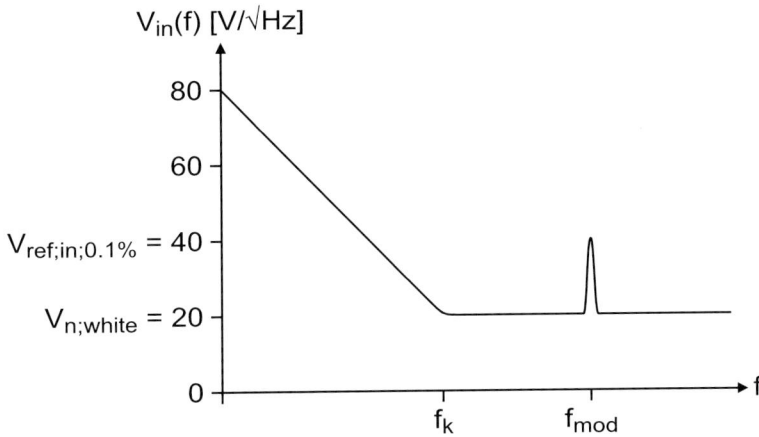

Figure 131. Spectral representation of the modulated reference signal

The noise root spectral density $V_n(f)$ is dominated, above the 1/f noise corner frequency f_k, by the white noise root spectral density $V_{n;white}$. It is precisely in this region that the reference signal is modulated, at a frequency $f_{mod} > f_k$. The 0.1 % input-referred fraction of the reference signal $V_{ref;in;0.1\%}$ emerges from the white noise floor $V_{n;white}$. In fact, it is twice higher. This allows its detection, but at the condition that the noise contribution is limited. Indeed, the total contribution of the white noise depends on the bandwidth. For the white noise, whose root spectral density is constant in function of the frequency, the total RMS amplitude $V_{n;RMS}$ in a bandwidth from f_{min} to f_{max} is:

$$V_{n;RMS} = \sqrt{\int_{f_{min}}^{f_{max}} V_n^2(f) df} = \sqrt{(f_{max} - f_{min}) V_{n;white}^2} \qquad (5.59)$$

To allow the accurate detection of the reference signal and a continuous compensation of the gain drift better than 0.1 %, the total input equivalent noise $V_{n;RMS}$ must not exceed $V_{ref;in;0.1\%}$:

$$V_{n;RMS} < V_{ref;in;0.1\%} \qquad (5.60)$$

Let's consider a symmetrical bandwidth around f_{mod} with a span f_{span}, i.e. in the frequency interval between

$$f_{min} = f_{mod} - \frac{f_{span}}{2} \qquad (5.61)$$

and

$$f_{min} = f_{mod} + \frac{f_{span}}{2} \qquad (5.62)$$

The total integrated RMS noise in this bandwidth is:

$$V_{n;RMS} = \sqrt{f_{span} V_{n;white}^2} = V_{n;white}\sqrt{f_{span}} \qquad (5.63)$$

Replacing equation 5.63 in 5.60 and using the values from table 17 gives:

$$f_{span} < \left(\frac{V_{ref;in;0.1\%}}{V_{n;white}}\right)^2 = \left(\frac{40}{20}\right)^2 = 4 \text{ [Hz]} \qquad (5.64)$$

To be able to extract the 0.1 % part of the reference signal $V_{ref;0.1\%}$ from the modulated preamplified signal V_{mod}, the bandwidth of the signal must be limited to 4 Hz, around a frequency f_{mod} in the order of 1 MHz. Figure 132 graphically represents these values.

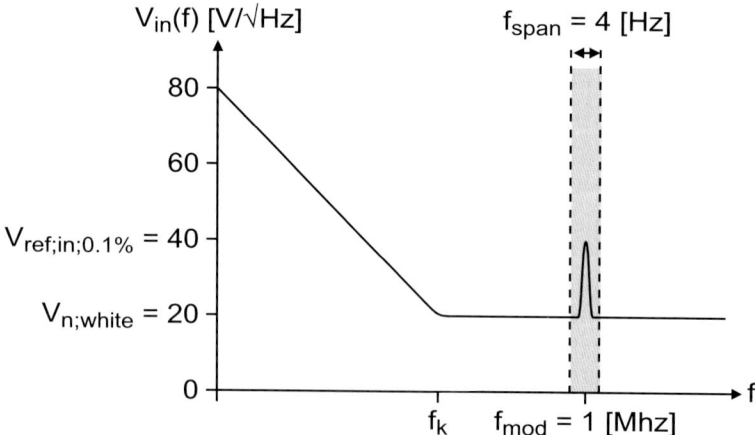

Figure 132. Band-limitation of the noise to increase the SNR

Clearly, a band-pass filter having these characteristics is not achievable. But if the signal is demodulated before band-limitation, i.e. brought back to DC, the noise contribution can be limited by *low-pass filtering*. This more advantageous situation is represented in figure 133.

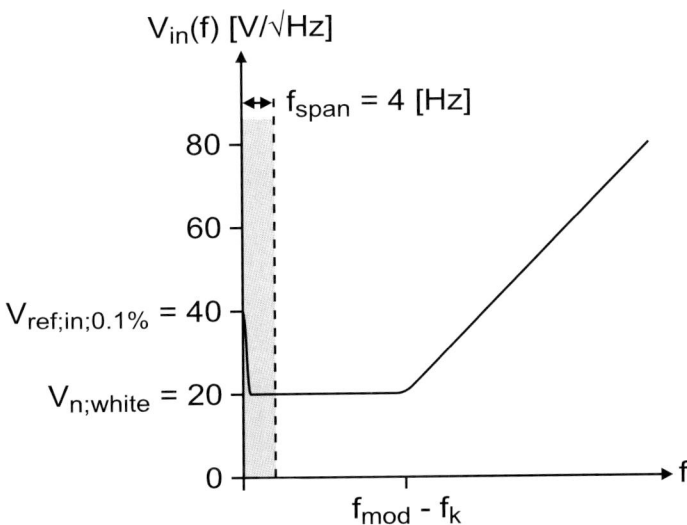

Figure 133. Low-pass filtering after demodulation to increase the SNR

If a low-pass filter with a cutoff frequency in the order of 1 Hz filters the output of the reference demodulator, it increases the signal-to-noise ratio to a sufficient value to allow the extraction of the 0.1 % part of the demodulated reference signal. It is noteworthy that the demodulated signal also contains a foldover noise component, due to the sampling process. It can be calculated using a similar development as in chapter 2, section 4.3.

In the system of figure 126, a second-order filtering is performed by the reference demodulator and the delta-sigma ADC. The demodulator performs a first-order low-pass filtering with a cutoff frequency $f_{p;demod}$ typically at 1 kHz. Then, the demodulated signal is filtered again (first-order) by the delta-sigma, with a cutoff frequency $f_{p;\Delta\Sigma}$ typically at 0.1 Hz. Figure 134 presents the resulting second-order low-pass filter transfer function, along with the partial transfer functions of the demodulator and delta-sigma.

The pole due to the demodulator transfer function is important mainly for the rejection of signal interferences (see section 6.8). However, it also improves the noise rejection for frequencies higher than $f_{p;demod}$, by making the global filter transfer function second-order.

The dominant pole is due to the delta-sigma transfer function. It limits the un-attenuated bandwidth to $f_{p;\Sigma\Delta} = 0.1$ Hz. The attenuation is then -20 dB/decade from 0.1 Hz to 1 kHz, and -40 dB/decade above 1 kHz (because of the pole of the demodulator).

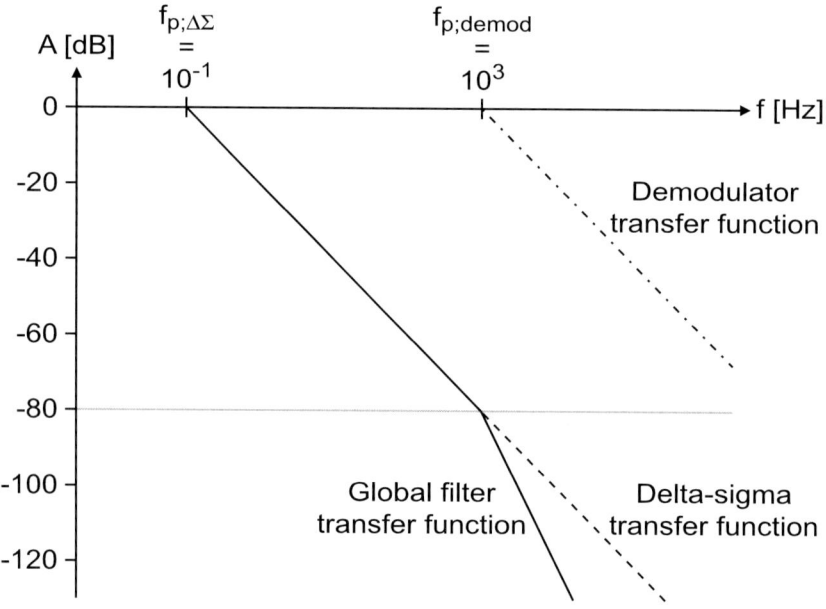

Figure 134. Demodulator and delta-sigma filter transfer functions

6.7 Delta-sigma analog-to-digital converter

The role of the delta-sigma converter in the system of figure 126 is to convert the DC demodulated reference signal $4V_{ref}$ into a digital value in order to implement the gain compensation feedback loop (section 6.4), but also to low-pass filter the noise (section 6.6).

The pole of the delta-sigma can easily be placed at very low frequency to implement the dominant cutoff of the low-pass filter transfer function discussed in section 6.6. This is because it is used in this system as an *analog-to-digital integrator*. This concept is explained below on the basis of figure 135, which presents the internal structure of the delta-sigma converter.

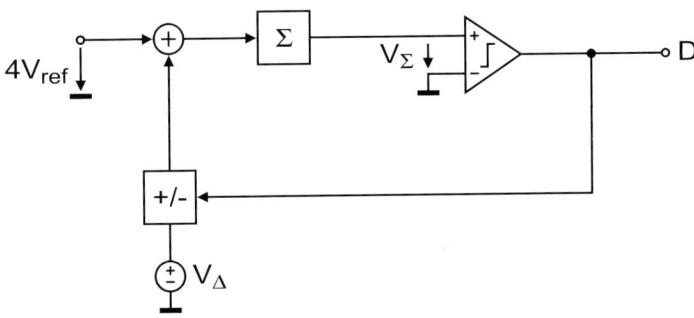

Figure 135. Delta-sigma used as an analog-to-digital integrator

The delta-sigma analog-to-digital converter is based on an analog integrator (Σ). To the integrated value, a difference signal V_Δ is periodically added or subtracted, while the input signal $4V_{ref}$ (to be converted) is added. The role of the difference signal is double: Firstly, since its sign is chosen in order to counterbalance the integrator value ($-V_\Delta$ if $V_\Sigma < 0$; $+V_\Delta$ if $V_\Sigma \leq 0$), it stabilizes V_Σ around zero. Secondly, if the number of positive (N_+) and negative (N_-) steps is memorized, their difference ($N_+ - N_-$) is the digitized image of the input signal.

In this case, the input is the reference demodulated signal $4V_{ref}$, which is generated every 4th demodulation phase. The delta-sigma thus needs to be synchronized with the demodulator. In fact, it is convenient to make them operate at the same frequency. In the delta-sigma, the signal from the demodulator is input every 4th phase, whereas the difference is added/subtracted once every phase in order to stabilize the integrator around zero. Figure 136 shows an example of the typical internal signals of the delta-sigma analog-to-digital converter of figure 135.

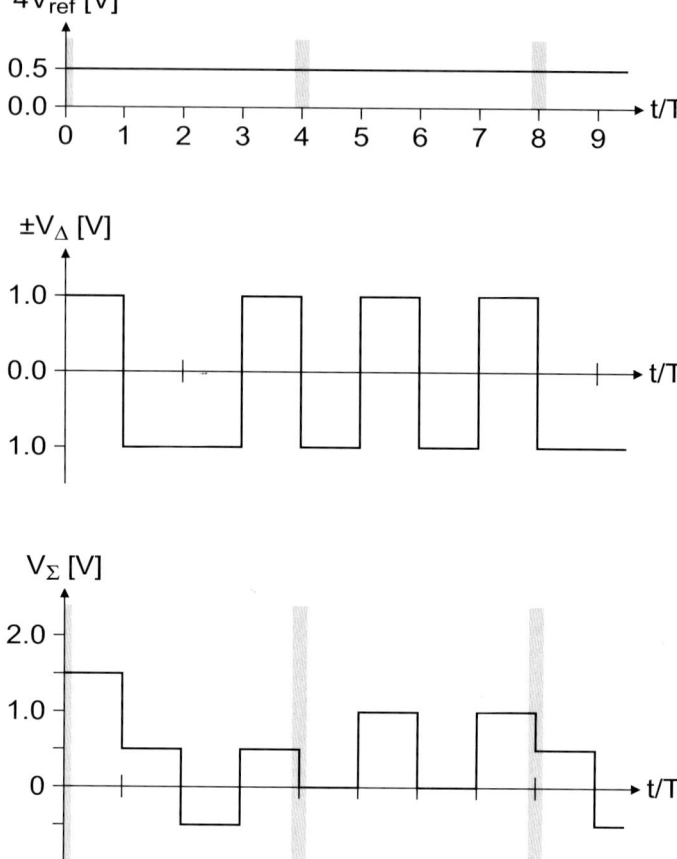

Figure 136. Typical signals in the delta-sigma modulator
($4V_{ref} = 0.5$ [V]; $V_\Delta = 1$ [V])

The period corresponding to one phase is T, and the x-axis in figure 136 is normalized to this value. Let's furthermore define the sampling instants t_i as integer multiples of T:

$$t_i = iT \qquad (5.65)$$

The constant input signal $4V_{ref}$ is added to the integrated value V_Σ every 4th phase, at the sampling instants t_i where $i = \{0, 4, 8, ...\}$. These input samplings are shaded in figure 136. Let's define:

Chapter 5: Hall microsystem with continuous calibration

$$V_{in;i} = \begin{cases} 4V_{ref} & i \bmod 4 = 0 \\ 0 & i \bmod 4 \neq 0 \end{cases} \quad (5.66)$$

$V_{in;i}$ represents the input voltage that is added to V_Σ at every sampling instant t_i. On the other hand, the difference $\pm V_\Delta$ is added at every t_i with i = {0, 1, 2, ...}. The increment at t_i is defined as:

$$V_{\Delta;i} = \pm V_\Delta \quad (5.67)$$

In equation 5.67, the sign is positive if the difference signal V_Δ is added, and negative if it is subtracted.

The initial value of V_Σ is arbitrarily 0. After the nth sampling instant, the output $V_{\Sigma;n}$ of the integrator is:

$$V_{\Sigma;n} = \sum_{i=0}^{n} V_{in;i} + V_{\Delta;i} \quad (5.68)$$

Let's name $N_{+;n}$ the total number of positive difference steps until the nth sampling instant, and $N_{-;n}$ the total number of negative steps. Let's furthermore suppose that n is an integer multiple of 4, and define

$$c = \frac{n}{4} \quad (5.69)$$

as the number of cycles of 4 phases. Using these definitions, equation 5.68 can be rewritten:

$$\begin{aligned} V_{\Sigma;n} &= \sum_{i=0}^{n} V_{in;i} + \sum_{i=0}^{n} V_{\Delta;i} = 4cV_{ref} + (N_{+;n} - N_{-;n})V_\Delta \\ &= nV_{ref} + (N_{+;n} - N_{-;n})V_\Delta \end{aligned} \quad (5.70)$$

The difference voltage V_Δ is chosen large enough to force V_Σ to remain in a given interval and avoid saturation:

$$-\varepsilon \leq V_{\Sigma;n} \leq \varepsilon \quad (5.71)$$

In this case, equation 5.70 becomes:

$$\varepsilon \geq |nV_{ref} + (N_{+;n} - N_{-;n})V_\Delta| \quad (5.72)$$

In other words, the difference between the summed input value and its estimation made with a difference of positive and negative steps is limited, and ε represents the precision of the estimation. The digital estimation of the input signal is:

$$D = N_{+;n} - N_{-;n} \cong -\frac{nV_{ref}}{V_\Delta} \quad (5.73)$$

To perform the analog-to-digital conversion, the integrator is used during a limited conversion time T_{conv} in the order of 1 s, corresponding to a conversion frequency of 1 Hz. On the other hand, the sampling period T is in the order of 1 µs, which corresponds to a frequency of 1 MHz.

For high frequencies, the sum of the input-related terms in equation 5.68 is equivalent to an integral and the delta-sigma modulator behaves as an integrator. For a periodic input signal $e^{j\omega t}$, the magnitude of the analog-to-digital transfer function can be approximated by:

$$A_{HF}(j\omega) = \left\| \int_0^{T_{conv}} e^{j\omega t} dt \right\| = \frac{1}{\omega} \quad (5.74)$$

On the other hand, for a DC signal, the input-related terms in the sum of equation 5.68 are all equal and correspond to the maximum possible gain value (normalized to a unity input signal):

$$A_{DC} = \left\| \int_0^{T_{conv}} 1 \, dt \right\| = T_{conv} = \frac{2\pi}{\omega_{conv}} \quad (5.75)$$

where ω_{conv} is the angular frequency corresponding to the conversion time T_{conv}. The high-frequency transfer function (equation 5.74) can be normalized to the maximum gain (equation 5.75):

$$A_{HF;norm}(j\omega) = \frac{A_{HF}(j\omega)}{A_{DC}} = \frac{\omega_{conv}}{2\pi\omega} \quad (5.76)$$

Equation 5.76 is the high-frequency asymptote of the signal transfer function. The low-frequency asymptote is simply:

Chapter 5: Hall microsystem with continuous calibration

$$A_{LF;norm}(j\omega) = \frac{A_{DC}}{A_{DC}} = 1 \quad (5.77)$$

The transfer function obtained is indeed that of a low-pass filter with a pole at:

$$\omega_{p;\Delta\Sigma} = \frac{\omega_{conv}}{2\pi} \quad (5.78)$$

The corresponding pole frequency $f_{p;\Delta\Sigma}$ is:

$$f_{p;\Delta\Sigma} = \frac{f_{conv}}{2\pi} \quad (5.79)$$

Figure 137 shows the plot of the low-pass transfer function resulting from the combination of both low- and high-frequency asymptotes, and the pole at their intersection.

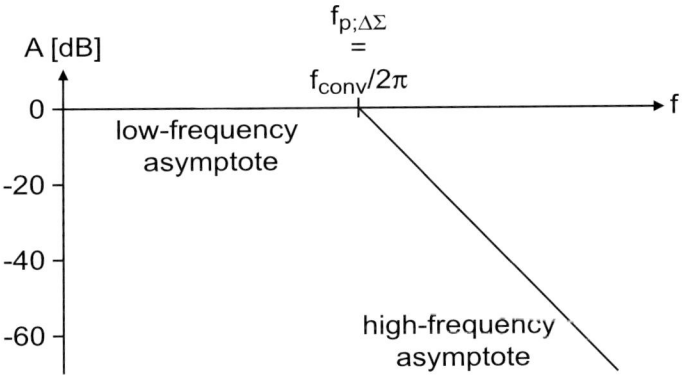

Figure 137. Low-pass filter function of the delta-sigma ADC

The low required cutoff frequency of 0.1 Hz can be obtained simply by using a sufficiently long conversion time $T_{conv} \cong 1.6$ s.

6.8 Rejection of signal interferences

Except for the noise discussed in section 6.6, a second problem motivates the use of a second-order low-pass filter in the gain adjustment feedback loop. Because the external component of the signal (V_{ext}) is not constant during the 4 demodulation phases of section 6.3, there is a parasitic component due to

V_{ext} in the reference demodulation signal. It is attenuated by the second-order low-pass filtering performed by the reference demodulator and delta-sigma.

The influence of the external component variation on the reference demodulation is presented on the grouped demodulation scheme of section 6.3.

If $V_{ext;1/2}$ and $V_{ext;3/4}$ are the mean values of the external signal components at the input of the reference demodulator during phases 1/2 and 2/3 respectively, equation 5.47 can be rewritten as:

$$V_{mod;1/2} - V_{mod;3/4} = 4V_{ref} + 2V_{ext;1/2} - 2V_{ext;3/4} \quad (5.80)$$

The demodulated signal contains the reference signal V_{ref}, but also a parasitic term $V_{ext;1/2} - V_{ext;3/4}$ depending on the variation of the external signal V_{ext} between phases 1/2 and 3/4. Let's name T_{mod} the modulation period, i.e. the duration of a phase, corresponding to a modulation frequency f_{mod} and an angular frequency ω_{mod}. Considering a periodic external signal $V_{ext} = e^{j\omega t}$ having a low frequency compared to the modulation frequency f_{mod}, the amplitude of the parasitic term (the low-frequency transfer function) in equation 5.80 can be calculated using the derivative:

$$A_{LF} = \|V_{ext;1/2} - V_{ext;3/4}\| \cong \left\|\frac{d}{dt}e^{j\omega t}\right\| 2T_{mod} = 2\omega T_{mod} \quad (5.81)$$

The maximum value of the differential term calculated by equation 5.81 cannot exceed 1, which corresponds to the normalized amplitude used to calculate the frequency influence. For high frequencies, the transfer function reaches its maximum:

$$A_{HF}(j\omega) = 1 \quad (5.82)$$

The function calculated by equation 5.81 is thus the low-frequency asymptote of a high-pass filter transfer function, whereas equation 5.82 is its high-frequency asymptote. The transfer function zero is located at:

$$\omega_Z = \frac{1}{2T_{mod}} = \frac{\omega_{mod}}{4\pi} \quad (5.83)$$

The corresponding zero frequency is:

$$f_z = \frac{f_{mod}}{4\pi} \tag{5.84}$$

Figure 138 shows the high-pass transfer function resulting from the combination of both low- and high-frequency asymptotes, and the zero at their intersection.

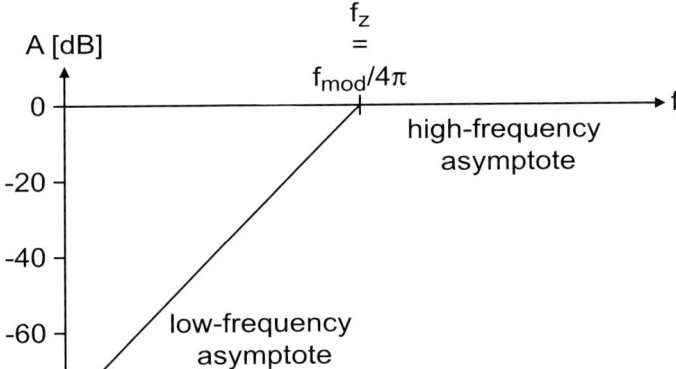

Figure 138. High-pass parasitic transfer function of the reference demodulator

A typical value for f_z is 100 kHz, corresponding to a modulation frequency f_{mod} of 1.26 MHz. If the alternate modulation/demodulation scheme presented in chapter 6, section 6.4 is used, the necessary modulation frequency corresponding to the same value of f_z is halved.

The parasitic term due to V_{ext} must still allow the correct extraction of the 0.1 % part of the reference signal in equation 5.80. For this reason, the filtered resulting level of V_{ext}, considering a maximum filter transfer function gain A_{max}, must not exceed the 0.1 % part of V_{ref}:

$$A_{max} V_{ext} \leq \frac{V_{ref}}{1000} \tag{5.85}$$

Because the signal level of V_{ext} is typically 100 times higher than the reference signal V_{ref}, equation 5.85 becomes:

$$A_{max} \leq \frac{1}{10^5} \tag{5.86}$$

In other words, the filtering must attenuate V_{ext} by at least $20 \log_{10}(10^5) =$ 100 dB in order to allow the 0.1 % extraction of the reference signal V_{ref} without interference from V_{ext}. Figure 139 presents the resulting parasitic transfer function. The V_{ext} signal parasitic transfer function and the second-order low-pass filtering performed by the reference demodulator and delta-sigma ADC are also plotted.

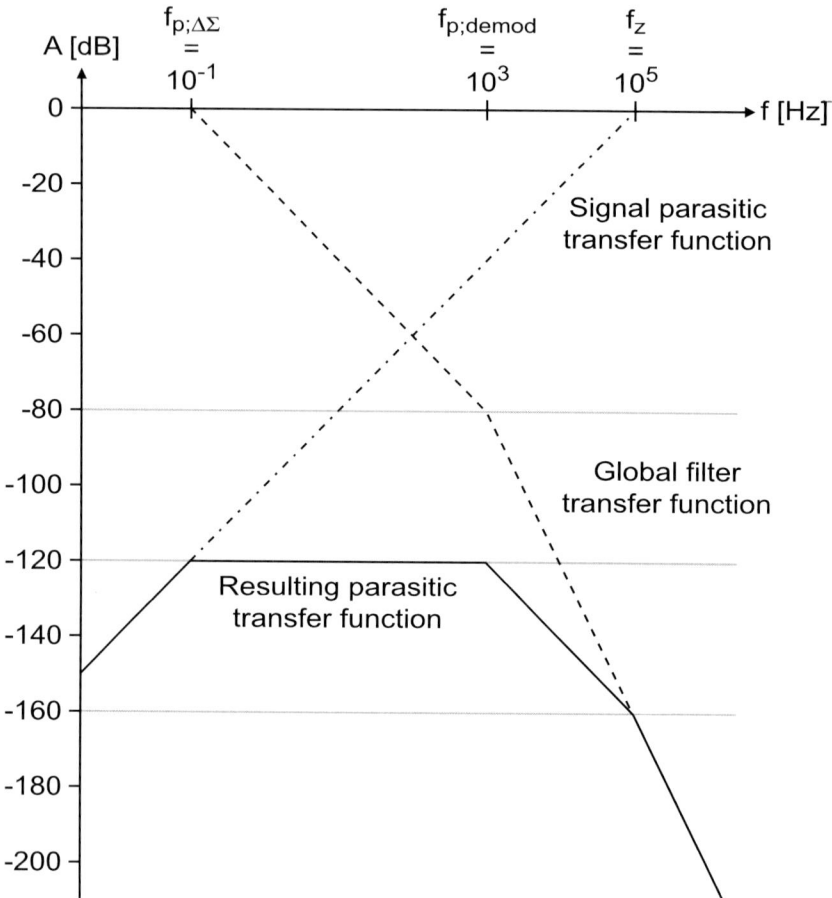

Figure 139. Parasitic transfer function before and after filtering

Thanks to the separation of the dominant pole $f_{p;\Delta\Sigma}$ of the delta-sigma and the parasitic zero f_z by 6 decades, the minimum attenuation is 120 dB. The demodulator pole helps to filter out high-frequency components with a second-order function, and to reduce the frequency span corresponding to the flat part of the resulting function. Because the minimum attenuation is 120 dB (20 dB higher than A_{max}), the filtering performed by the reference demodulator and the delta-sigma converter sufficiently limits the parasitic signal component (due to the variation of V_{ext}) to allow the extraction of the 0.1 % part of the reference signal V_{ref}.

7   CONCLUSION

The calibration system presented in this chapter allows to continuously adjust the sensitivity of a Hall sensor microsystem, without interrupting normal operation. Because it operates at high frequency, the circuit can be used in applications requiring a large bandwidth, for instance in contact-less current measurement systems. The implemented gain and offset compensations are inspired from both chopper and autozero techniques. They are typical applications of the compensation methodology presented in chapter 4 and of the associated correction circuits.

Chapter 6

Implementation of the Hall microsystem with continuous calibration

This chapter presents the implementation of the current measurement Hall microsystem with continuous digital gain calibration described in chapter 5. The issues and solutions at block-level are detailed and discussed on the basis of simulation and measurement results. System-level imperfections are also presented, and compensation techniques proposed. The complete ASIC is then described. Its features and measurement results are discussed. Finally, future development possibilities are proposed.

1 INTRODUCTION

A prototype of the Hall microsystem presented in chapter 5 (figure 126) [91][92] has been realized in a conventional 0.8 µm 5V CMOS technology. It has voluntarily been designed for test and validation purposes [96]. To this end, the circuit allows *maximum test and configuration possibilities*, and on some design parameters the flexibility is favoured at the cost of pure performance. The circuit is also kept as modular and simple as possible to allow easy identification of problems and limitations.

This chapter presents the implementation issues at block- and system-level, and details the digital compensation techniques that can be used to correct the imperfections. Both chopper- and autozero-like techniques (chapter 2) are implemented. This current measurement ASIC is a typical application for the digital compensation methodology of chapter 4 and the associated correction circuits (chapter 3).

2 HALL SENSOR ARRAY

Instead of a single Hall sensor, an array of 4 Hall sensors is used to measure the magnetic field. This topology increases the signal-to-noise ratio and minimizes the offset [81]. To generate the reference magnetic field, 4 coils are also integrated, one above each sensor [80]. Figure 140 shows the array of sensors and coils.

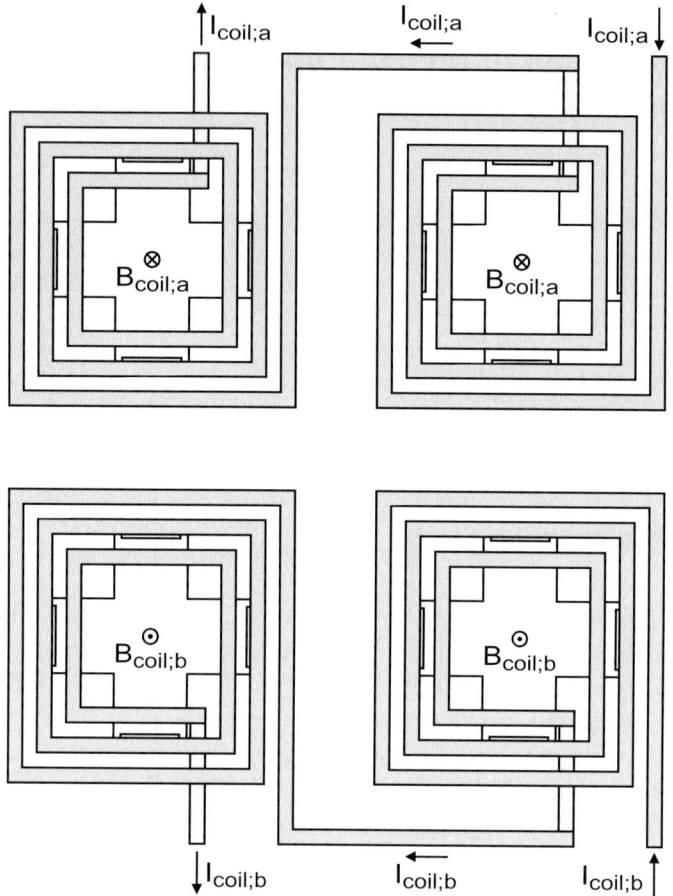

Figure 140. Hall sensor and reference coil array

The coils are connected horizontally by pairs. When both $I_{coil;a}$ and $I_{coil;b}$ are positive, the reference magnetic field $B_{coil;a}$ for the two top sensors is in the opposite direction as the field $B_{coil;b}$ for the two bottom sensors. This allows, for one of the alternate modulation schemes presented in section 6.4, to cancel the external field and measure only the reference signal. The cancellation is achieved by subtracting the measurements of both pairs of sensors. If the current direction in one of the coil pairs is reversed, it is also possible to generate a unidirectional reference field that is added by the 4 sensors. The

Chapter 6: Implementation of the Hall microsystem

latter configuration corresponds to the circuit described in chapter 5 and is used in the microsystem.

Table 18 presents the main characteristics of the Hall sensors and the coils (1 element of the array). These values are also used in chapter 5.

Table 18. Sensor and coil characteristics

	Parameter	Value	Unit
Sensor	Current-related sensitivity	80	V/TA
	Bias current	1	mA
	Sensitivity	80	mV/T
Coil	Efficiency	0.125	T/A
	Bias current	4	mA
	Magnetic field	0.5	mT

3 PREAMPLIFIER

The preamplifier generates an amplified modulated signal V_{mod}, which is fed into the parallel demodulators for the external signal (V_{ext}), the reference (V_{ref}) and the offset (V_{off}). It constitutes the interface between the sensors and the signal processing circuits.

3.1 Programmable gain range preamplifier

Figure 141 presents the block diagram of the preamplifier. It is constituted of 3 stages.

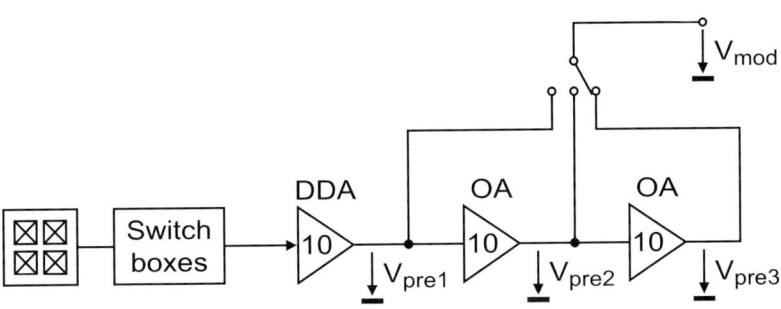

Figure 141. Preamplifier block diagram

The first stage consists in a Differential Differences Amplifier (DDA), which amplifies the signals of the 4 sensors composing the array jointly. The signal is modulated (see chapter 5, section 6.2) at a frequency of 1 MHz. Two more operational amplifiers (OA) are chained, and the gain of each stage is fixed to 10 (20 dB). The signal after 1, 2, or 3 stages is available as input for any of the demodulators presented in section 4, thanks to a configurable crossbar switch (only one row displayed in figure 141). For each demodulator, the preamplification gain can thus separately be set to 20, 40 or 60 dB.

3.2   DDA

Since the input signal level is low (< 4 mV) and the sensor output impedance is in the order of 500 Ω, the main requirements of the first stage of the preamplifier are a low noise level and high input impedance. For these reasons, a DDA is chosen [97]. Figure 142 presents a schematic of the first stage of the preamplifier and its connection to the sensor array.

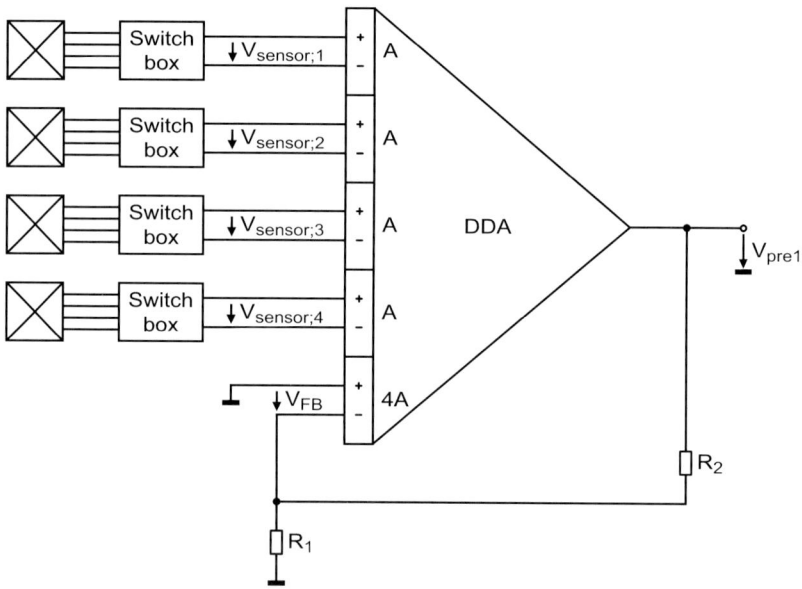

Figure 142.   Sensor array and first stage of the preamplifier

Each Hall sensor is connected to a separate differential input of the DDA, through a switch box implementing the spinning current technique presented in chapter 5, section 3. An additional input is used for feedback to stabilize the gain of the first stage at 20 dB. The differential inputs are realized by differential pairs connected in parallel, providing a high input impedance. The open-loop gains of the 4 sensor inputs are equal (to A), whereas the relative gain of the feedback differential input is 4 times larger (4A). Figure 143 presents a model of the DDA with its feedback loop.

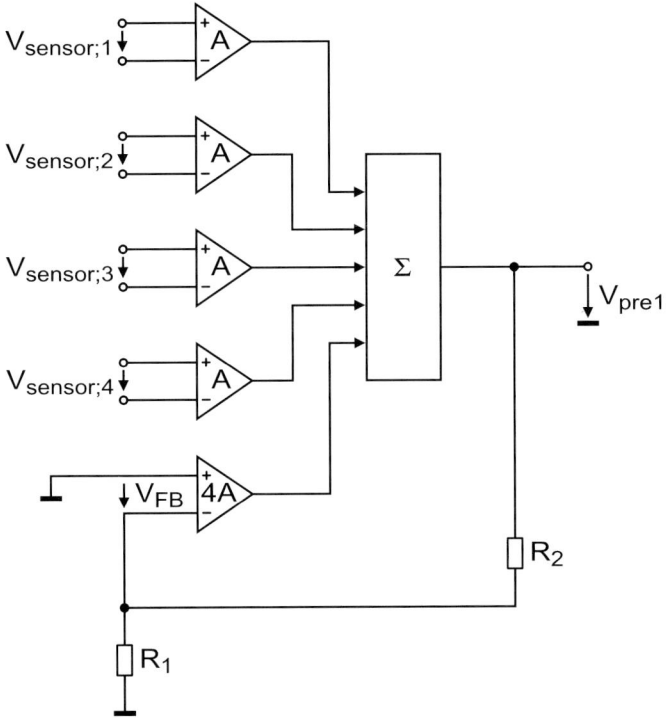

Figure 143. Model of the DDA with 5 differential inputs

The output signal of the DDA is the sum of the amplified differences (hence the name "Differential Differences Amplifier"):

$$V_{pre1} = A(V_{sensor;1} + V_{sensor;2} + V_{sensor;3} + V_{sensor;4} + 4V_{FB}) \qquad (6.1)$$

The resistive ladder generates the feedback voltage V_{FB}:

$$V_{FB} = -V_{pre1}\frac{R_1}{R_1 + R_2} \qquad (6.2)$$

Replacing equation 6.2 in 6.1 gives:

$$V_{pre1} = \frac{1}{4\dfrac{R_1}{R_1+R_2} + \dfrac{1}{A}}(V_{sensor;1} + V_{sensor;2} + V_{sensor;3} + V_{sensor;4}) \quad (6.3)$$

If the open-loop gain A is infinite, equation 6.3 becomes:

$$V_{pre1} = \frac{R_1 + R_2}{R_1} \cdot \frac{V_{sensor;1} + V_{sensor;2} + V_{sensor;3} + V_{sensor;4}}{4} \quad (6.4)$$

This result is similar to the one obtained for a simple operational amplifier (equations 4.9), except that in this case the sum of all the differential inputs is amplified. In fact, since the gain of the feedback differential input is 4 times higher and thus corresponds to the cumulated gain of the 4 remaining inputs, the DDA amplifies the mean V_{sensor} of the 4 differential input voltages:

$$V_{sensor} = \frac{V_{sensor;1} + V_{sensor;2} + V_{sensor;3} + V_{sensor;4}}{4} \quad (6.5)$$

Using this definition, equation 6.4 indeed becomes:

$$V_{pre1} = \frac{R_1 + R_2}{R_1} V_{sensor} \quad (6.6)$$

By choosing $R_2 = 9R_1$, the gain of the first preamplifier stage referred to the mean sensor voltage[1] is set to:

$$A_{pre1} = \frac{R_1 + R_2}{R_1} = 10 \quad (6.7)$$

Figure 144 presents the schematic of the DDA. It uses a folded-cascode topology, with a single-output class A Miller-compensated second stage.

The gain of the feedback input (V_{FB+}; V_{FB-}) is made 4 times higher than for the 4 other inputs (V_{i+}; V_{i-}, $i \in [1, 4]$). This is done by multiplying the bias current of the corresponding differential pair by 4, and by using an aspect ratio 4W/L in the feedback differential pair instead of W/L (as for the other inputs). In this way, the feedback input behaves equivalently to the sum of the 4 signal inputs, and equation 6.6 applies when feedback is used.

---

[1] or equivalently: to a single sensor

Chapter 6: Implementation of the Hall microsystem   205

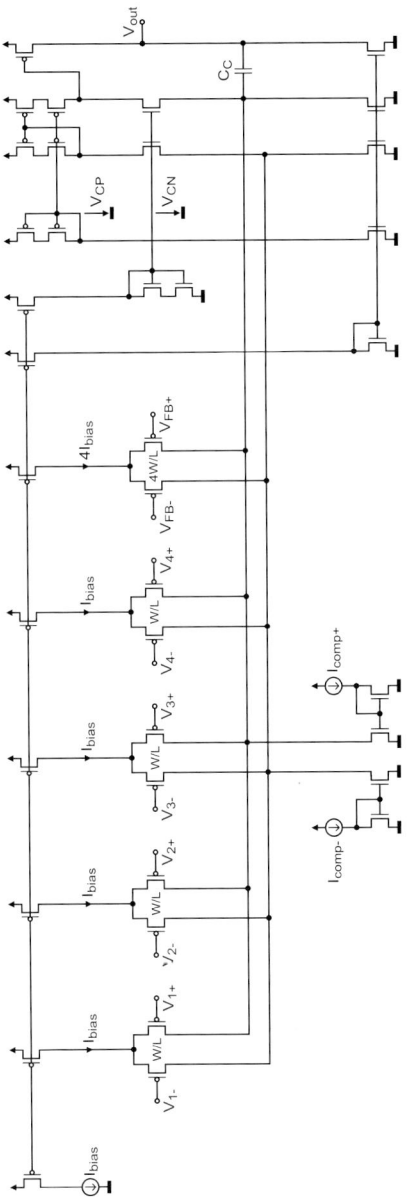

Figure 144. Schematic of the DDA

An offset compensation is performed (see chapter 4) by injecting two compensation currents I_{comp+} and I_{comp-} in differential mode. The compensation nodes are chosen so that the effect of the compensation current mirrors on other circuit characteristics is negligible and symmetrical. In fact, the differential compensation current is simply added to folding current generated by the NMOS bias transistors (current mirror). An alternate differential compensation technique can be found in [98], and a single-ended compensation is also possible [51].

As in Miller amplifiers (see chapter 4, section 2), the compensation capacitor C_C sets the dominant pole. The series resistor is here directly the NMOS cascode transistor. The NMOS and PMOS cascode voltages V_{CN} and V_{CP} respectively are generated internally.

Table 19 presents the main characteristics [7] of the DDA, for a capacitive output load of 2 pF in parallel with a 45 kΩ. resistive load, at room temperature (T = 27 °C).

Table 19. Characteristics of the DDA

Parameter	Value	Unit
Supply voltage	5	V
Supply current	0.8	mA
Temperature range	-40 to 125	°C
Phase margin	70	°
Open-loop gain (signal input)	78	dB
Open-loop gain (feedback input)	90	dB
GBW	30	MHz
Slew rate	2.5	V/μs
Input-referred offset (3σ)	± 8.5	mV
Offset compensation sensitivity	0.5	mV/μA
Input-referred noise (@ 1 MHz)	15	nV$\sqrt{\text{Hz}}$
Signal inputs common-mode range	0.5 to 2.5	V
Nonlinearity	< 0.1	%
Negative power supply rejection ratio (PSRR$^-$)	95	dB
Positive power supply rejection ratio (PSRR$^+$)	85	dB

Special attention is devoted to the linearity of the signal input differential pairs. Since there is no feedback on the 4 signal input pairs, they must indeed have a high intrinsic linearity. Furthermore, they must provide an extended input common-mode range. This is necessary because the sensor resistance, and thus its common-mode voltage, varies with temperature.

A high bandwidth of 30 MHz and an open-loop gain of 90 dB are achieved at the sacrifice of a higher power consumption. The high bandwidth is necessary to allow the amplification of signals at 1 MHz (the spinning frequency) with a closed-loop gain of 20 dB.

3.3 Operational amplifier

The operational amplifier presented in this section is used in the preamplifier, but also in the demodulator and as output buffer. The only difference between the different instances of the amplifier is the compensation capacitor, which is adjusted as a function of the capacitive load.

Figure 145 presents the schematic of the operational amplifier. It uses the same folded-cascode topology and single-output Miller second stage as the DDA presented in section 3.2.

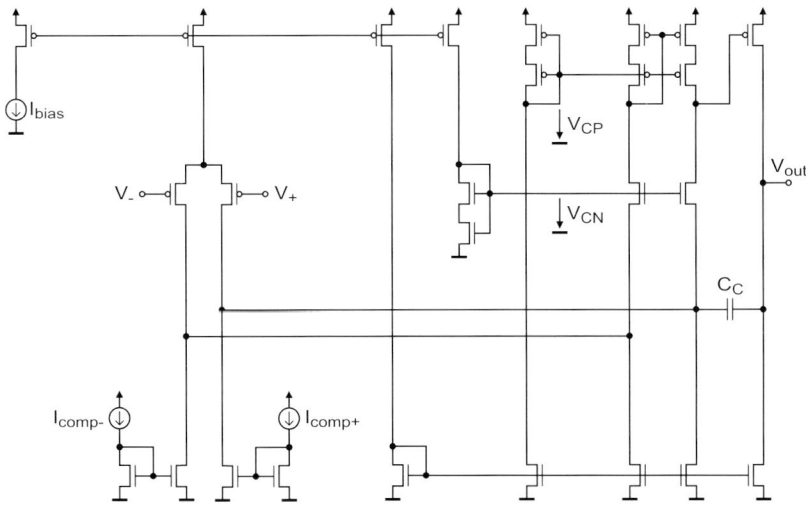

Figure 145. Schematic of the operational amplifier

Table 20 presents the main characteristics of the operational amplifier, for a capacitive output load of 2 pF in parallel with a 45 kΩ. resistive load, at room temperature (T = 27 °C).

Since this amplifier is used in switched-capacitor circuits and as second and third preamplifier, it has a high slew rate. This guarantees a rapid output voltage swing even for large signal amplitudes. Combined with the high bandwidth, this also ensures a short settling time.

Table 20. Characteristics of the operational amplifier

Parameter	Value	Unit
Supply voltage	5	V
Supply current	1.0	mA
Temperature range	-40 to 125	°C
Phase margin	70	°
Open-loop gain	100	dB
GBW	45	MHz
Slew rate	20	V/µs
Input-referred offset (3σ)	± 8.7	mV
Offset compensation sensitivity	1	mV/µA
Input-referred noise (@ 1 MHz)	15	nV√Hz
Input common-mode range	0.5 to 4	V
Nonlinearity	< 0.1	%
Negative power supply rejection ratio ($PSRR^-$)	100	dB
Positive power supply rejection ratio ($PSRR^+$)	85	dB

4 DEMODULATORS

The operational amplifier presented in section 3.3 is used to implement switched-capacitor demodulator circuits. Each external signal, reference and offset demodulator has its own characteristics and optimization possibilities. They are presented separately in the sections below, after an introduction to switched-capacitor integrators.

4.1 Switched-capacitor integrators

Operational amplifier-based switched-capacitor circuits are a simple yet powerful means of implementing the synchronous demodulators required in the current measurement microsystem [69]. A complete tutorial about switched-capacitor circuits and their design can be found in [7]. The focus is set here on the application to synchronous demodulators, and the associated switched-capacitor integrator circuit presented in figure 146.

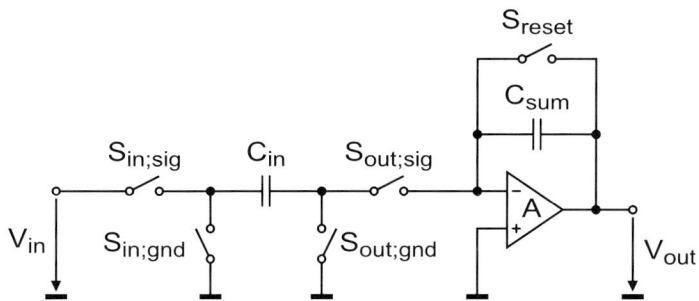

Figure 146. Switched-capacitor integrator

The input signal V_{in} is fed to the virtual ground of the operational amplifier through an input capacitor C_{in} associated to 4 switches $S_{in;sig}$, $S_{in;gnd}$, $S_{out;gnd}$ and $S_{out;sig}$. A summing capacitor C_{sum} is connected on the feedback path of the amplifier, which thus acts as an integrator: C_{sum} integrates the charges input through switch $S_{out;sig}$ into the virtual ground created by the amplifier. The output voltage is:

$$V_{out} = \frac{Q_{sum}}{C_{sum}} \qquad (6.8)$$

where Q_{sum} is the total charge in C_{sum}. A reset switch S_{reset} allows to remove all the charges ($Q_{sum} = 0$) from C_{sum} and to consequently reset the integrator ($V_{out} = 0$).

If the 4 switches of the input block are operated in two successive steps, as presented in figure 147, V_{out} increases by the input voltage value $V_{in;i}$ multiplied by the C_{in}/C_{sum} ratio. It is assumed that $V_{in;i}$ remains constant during both steps, and that the initial output voltage is $V_{out;i-1}$, corresponding to a charge $Q_{sum;i-1}$ in C_{sum}.

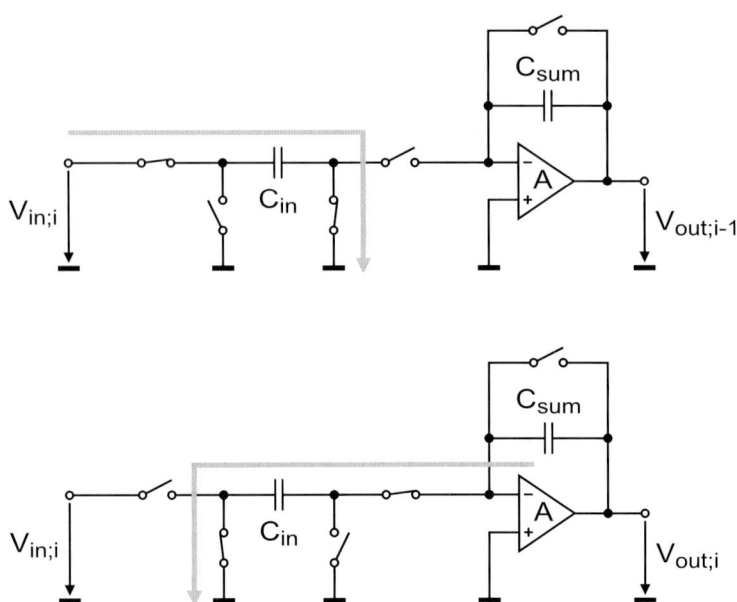

Figure 147. Addition principle
Top: First step (input sampling); *Bottom*: Second step (output update)

The first step is the *input sampling*. The input voltage is sampled on the input capacitor C_{in} through $S_{in;sig}$ and $S_{out;gnd}$. The corresponding stored charge $Q_{in;i}$ is:

$$Q_{in;i} = C_{in} V_{in;i} \qquad (6.9)$$

The switches are then all open before the transition to the second step, which performs the *output update*. Because the input capacitor is connected to the ground through $S_{in;gnd}$ and to the virtual ground through $S_{out;sig}$, the charge $Q_{in;i}$ is completely transferred to the output capacitor. The total charge $Q_{sum;i}$ in C_{sum} becomes:

$$Q_{sum;i} = Q_{sum;i-1} + Q_{in;i} \qquad (6.10)$$

The output voltage thus becomes:

$$V_{out;i} = \frac{Q_{sum;i}}{C_{sum}} = \frac{Q_{sum;i-1} + Q_{in;i}}{C_{sum}} = V_{out;i-1} + V_{in;i} \frac{C_{in}}{C_{sum}} \qquad (6.11)$$

The input voltage $V_{in;i}$ is amplified by the ratio of the input and summing capacitor, and added to the previous output voltage $V_{out;i-1}$. The summing capacitor is thus indeed an accumulator (or integrator).

With the same principle, it is also possible to subtract the input voltage $V_{in;i}$, by using the two successive steps presented in figure 148.

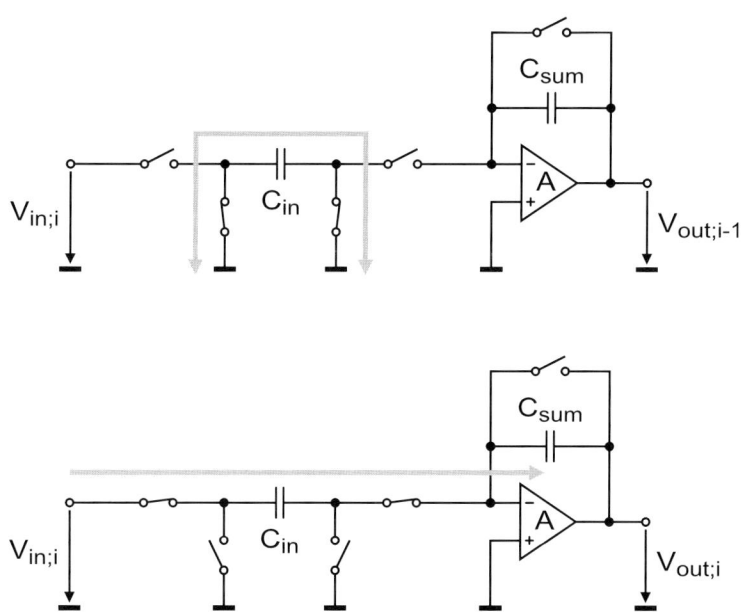

Figure 148. Subtraction principle
Top: First step (preparation); Bottom: Second step (output update)

During the *preparation* step, the input capacitor is discharged by short-circuiting its terminals to ground through switches $S_{in;gnd}$ and $S_{out;gnd}$. During the second step, the input capacitor is charged through $S_{in;sig}$ and $S_{out;sig}$. The corresponding charge $Q_{in;i}$ is:

$$Q_{in;i} = C_{in} V_{in;i} \qquad (6.12)$$

Because the virtual ground is at high impedance, this charge is in fact extracted from the summing capacitor. The total charge $Q_{sum;i}$ in C_{sum} thus becomes:

$$Q_{sum;i} = Q_{sum;i-1} - Q_{in;i} \qquad (6.13)$$

Consequently, the output voltage becomes:

$$V_{out;i} = \frac{Q_{sum;i}}{C_{sum}} = \frac{Q_{sum;i-1} - Q_{in;i}}{C_{sum}} = V_{out;i-1} - V_{in;i}\frac{C_{in}}{C_{sum}} \qquad (6.14)$$

The signals $\phi_{in;sig}$, $\phi_{in;gnd}$, $\phi_{out;gnd}$ and $\phi_{out;sig}$ control the switches $S_{in;sig}$, $S_{in;gnd}$, $S_{out;gnd}$ and $S_{out;sig}$ respectively. Figure 149 presents the necessary timing to implement the two steps for the addition of figure 147. A high logic level corresponds to a closed switch, whereas a low logic level causes the switch to be open.

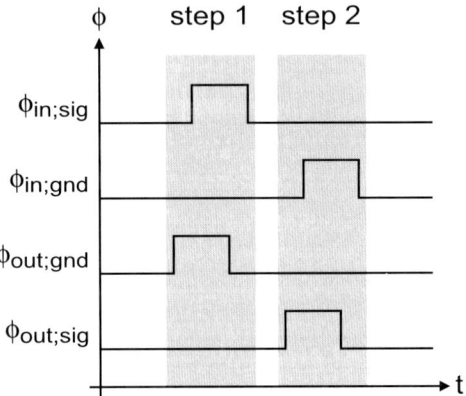

Figure 149. Switch timing for an addition

Figure 150 presents the necessary timing to implement the two steps for the subtraction of figure 148.

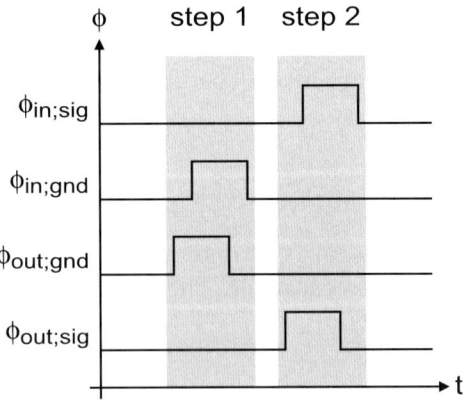

Figure 150. Switch timing for a subtraction

The clocks are non-overlapping to guarantee that the transition from one step to another does not cause charge losses. Furthermore, delayed clock signals [7] are used to make the system insensitive [99] to parasitic capacitances [100]. By always opening first the switch connected to ground (or to the virtual ground), signal-dependent charge injection [8] is avoided. The signal-independent charge injection simply causes an offset.

Another imperfection of the system is due to the finite bandwidth of the amplifier. Its effect can be reduced by using a step frequency inferior by at least one decade to the bandwidth of the amplifier [7]. This is the case in the current measurement microsystem if the operational amplifier of section 3.3 (bandwidth of 45 MHz) is used at a frequency of 1 MHz.

Finally, the finite gain of the amplifier has also an influence on the characteristics of the integrator [9][101]. It causes mainly a magnitude error, as in continuous-time systems.

The 3 demodulators for the external, reference and offset signals are implemented using this circuit. The capacitors are implemented using two superposed polysilicon layers (poly-poly capacitors). However, in fully digital technologies, they can be implemented using the parasitic capacitances of MOS transistors [102] (MOScaps).

4.2   External signal demodulator

The external signal demodulator uses the switched-capacitor integrator presented in section 4.1, using equal values for the input and summing capacitor ($C_{in} = C_{sum} = 2$ pF). For each of the 4 demodulation phases of table 16, the timing corresponding to the appropriate operation (addition or subtraction) is generated. Figure 151 shows the timing corresponding to the +-+- sequence necessary to extract $4V_{ext}$ from V_{mod}.

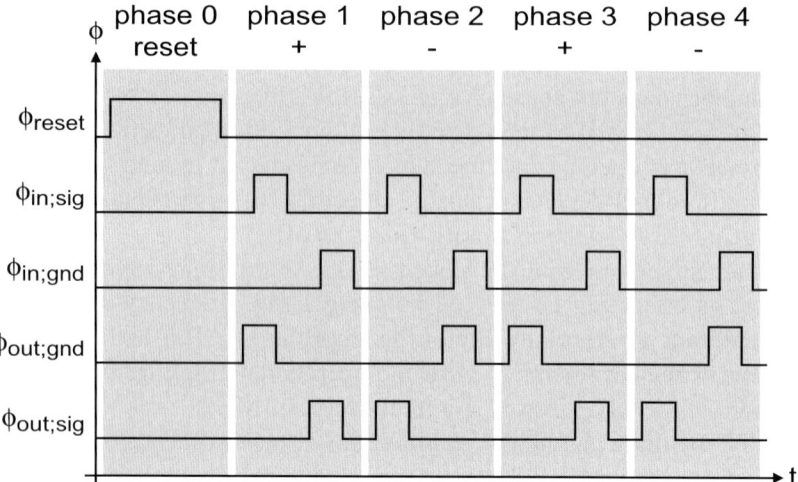

Figure 151. External signal demodulator switch timing

First, the output is reset using the control signal ϕ_{reset} of the S_{reset} switch. Then, the 4 operation phases alternate additions and subtractions.

It is noteworthy that the reset phase in fact does not add an additional phase, since it can be included in one of the 4 demodulation phases, at a step where switch $S_{out;sig}$ is open. The timing presented here is simply used for clarity.

The timing for the + operations is generated as presented in figure 149. On the other hand, the timing for - operations slightly differs from the one of figure 150. It is modified because the demodulator must be synchronized with the modulated output of the preamplifier. In particular, it is important to sample the modulated signal after a *fixed time interval* from the beginning of the phase. But on the timing of figure 150, the input signal is sampled during step 2 (instead of step 1 as for the addition of figure 149). To solve this problem, both steps 1 and 2 are simply exchanged for the subtraction. This is possible since the last performed step before a subtraction (the second step of an addition) causes the capacitor C_{in} to be discharged, exactly as during the first step of a subtraction according to figure 150. It is thus possible to suppress the first step of the subtraction, or exchange both steps as in figure 151. In the latter case, additions and subtractions can be chained in any order.

Another important point that must be considered about the synchronization of the demodulator with the modulated output of the preamplifier is the *phase shift*. In fact, the beginning of the demodulation phase must be shifted by half a period with respect to the modulation phase, as shown in figure 152.

By doing so, the preamplifier has more time to settle and thanks to the technique presented in figure 151, the modulated signal is always sampled after a sufficient and constant settling time. In figure 152, the modulation and demodulation phases are shown in light grey, and the demodulation samplings in dark grey.

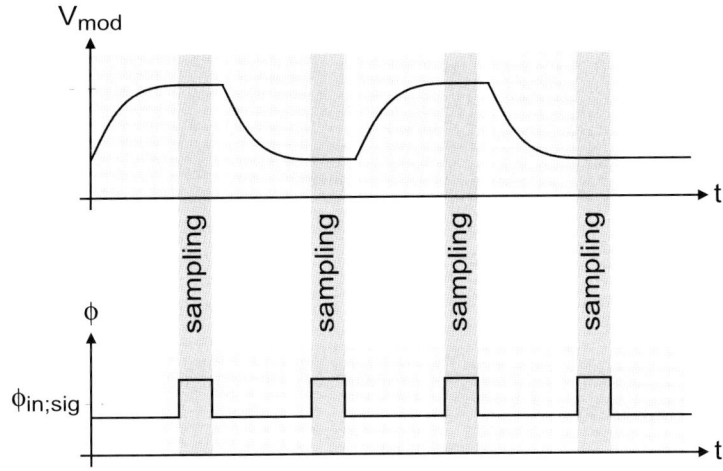

Figure 152. Demodulator phase shift

During the demodulation cycle, the integrator performs a step-by-step calculation of the addition/subtraction of the 4 successive modulated values $V_{mod;i}$. Since the system is not reset between the phases, the final voltage $V_{out;i}$ of phase i (i ∈ [0, 4]) is also the initial voltage of the next phase, and thus:

$$V_{out;i} = V_{out;i-1} + sgn_i V_{mod;i} = V_{out;0} + \sum_{j=1}^{i} sgn_j V_{mod;j} \qquad (6.15)$$

where sgn_i represents the sign of the operation performed during phase i and is defined as:

$$sgn_i = \begin{cases} 1 & \text{addition} \\ -1 & \text{subtraction} \end{cases} \qquad (6.16)$$

In equation 6.15, $V_{mod;i}$ is the value during phase i of the modulated output signal of the preamplifier. Furthermore, because the system is reset before phase 1, $V_{out;0} = 0$. Table 21 shows how the final demodulated result is obtained, by presenting the intermediate output voltage values of the external signal demodulator according to equation 6.15 and figure 151.

Table 21. External signal demodulation intermediate results

Phase (i)	sgn_i	$V_{out;i}$
0	-	0
1	1	$V_{mod;1}$
2	-1	$V_{mod;1} - V_{mod;2}$
3	1	$V_{mod;1} - V_{mod;2} + V_{mod;3}$
4	-1	$V_{mod;1} - V_{mod;2} + V_{mod;3} - V_{mod;4}$

At the end of the 4th phase, the output of the demodulator corresponds to the result of equation 5.41:

$$V_{out;4} = 4V_{ext} \qquad (6.17)$$

This result can be sampled and held by an additional stage while the signal demodulator performs the next demodulation cycle.

4.3 Reference demodulator

The reference demodulator is implemented on the basis of the signal demodulator, with two additional blocks.

The first addition is a second input path for a coarse reference voltage V_{Cref} allowing to roughly set the mean output voltage (DC level) of the reference demodulator output to null. In this way, the necessary full scale of the delta-sigma analog-to-digital converter is reduced, which allows to increase the resolution of the gain measurement.

The second addition is a feedback path, which transforms the integrator into a low-pass filter. This is important to reduce the noise (chapter 5, section 6.6) and external interferences levels (chapter 5, section 6.8).

Figure 153 presents the reference demodulator obtained after the addition of these two elements.

Chapter 6: Implementation of the Hall microsystem

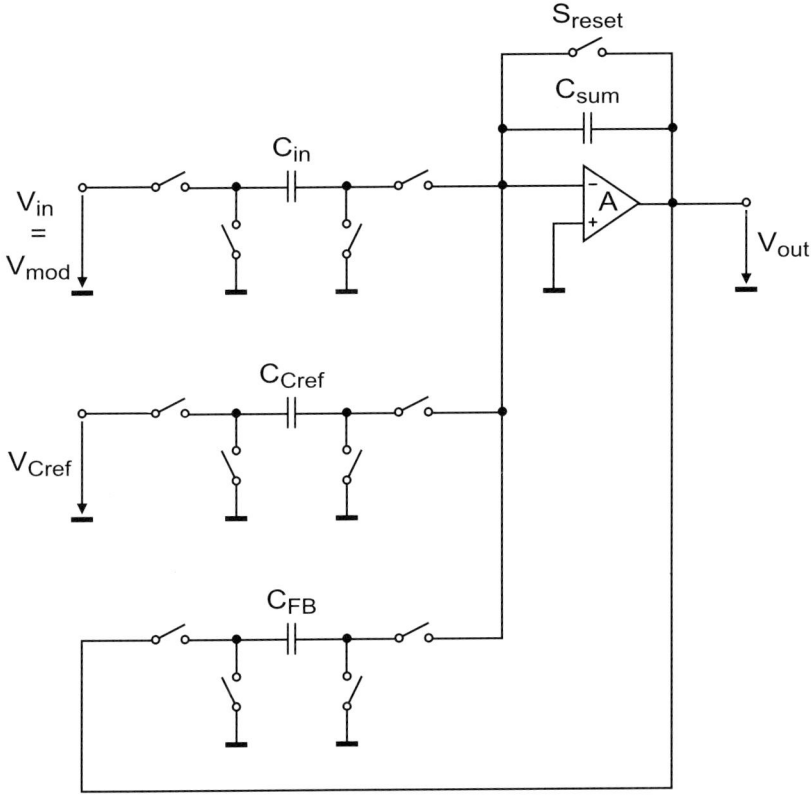

Figure 153. Reference demodulator

Figure 154 shows the switch timing used for the reference demodulator presented in figure 153.

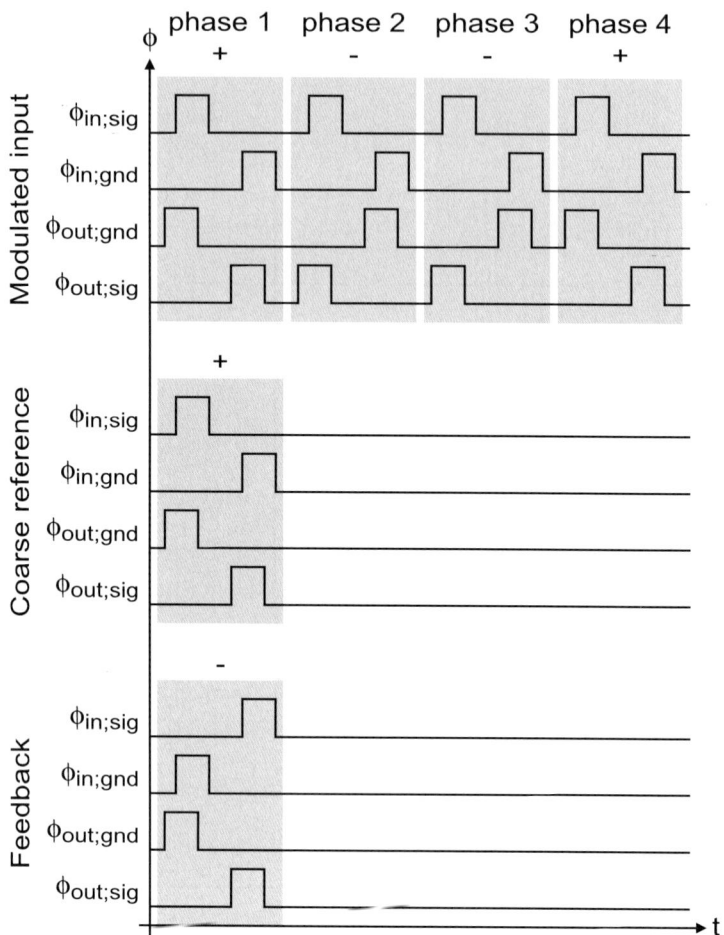

Figure 154. Reference signal demodulator switch timing

The reset switch is not used here but can, however, be activated one single time at power-up. The 4 switches connected to the modulated input ($V_{in} = V_{mod}$) and associated to the C_{mod} capacitor implement the timing for the +--+ sequence necessary to extract $4V_{ref}$ from the input V_{mod} (see table 16).

Once per cycle, during the first of 4 phases, the reference voltage V_{Cref} is subtracted. This coarse voltage is chosen to roughly cancel the $4V_{ref}$ component at the output:

$$V_{Cref} \cong 4V_{ref}\frac{C_{in}}{C_{Cref}} \qquad (6.18)$$

Chapter 6: Implementation of the Hall microsystem

It is noteworthy that several operations can be performed simultaneously, as during the second step of phase 1. The sum of the charges in the corresponding capacitors are then transferred to the summing capacitor C_{sum} in one single step.

In addition, this property is used to feed back a fraction of the output V_{out} using the feedback path through C_{FB} during the first phase. This causes the circuit to behave as a low-pass filter. It can be shown that the z-domain transfer function (from the signal input V_{in} to the output V_{out}) of the circuit is [7]:

$$H(z) = \frac{\alpha_1 z^{-1}}{1 + \alpha_2 - z^{-1}} \qquad (6.19)$$

with

$$\alpha_1 = \frac{C_{in}}{C_{sum}} \qquad (6.20)$$

and

$$\alpha_2 = \frac{C_{FB}}{4 C_{sum}} \qquad (6.21)$$

The factor 4 in equation 6.21 is caused by the fact that the feedback is performed every 4th phase only.

Equation 6.19 can be translated in the s-domain, using:

$$z^{-1} = e^{-sT_{mod}} \cong 1 - sT_{mod} \qquad (6.22)$$

The right part of equation 6.22 is valid for low input frequencies with respect to the modulation frequency f_{mod} (corresponding to a phase period T_{mod}). Under this assumption, the term sT_{mod} is very small and:

$$z^{-1} \cong 1 \qquad (6.23)$$

Replacing equations 6.22 and 6.23 into equation 6.19 gives:

$$H(s) \cong \frac{\alpha_1}{\alpha_2 + sT_{cmod}} = \frac{\frac{\alpha_1}{\alpha_2}}{1 + s\frac{T_{mod}}{\alpha_2}} = \frac{A}{1 + \frac{s}{\omega_p}} \qquad (6.24)$$

Equation 6.24 is indeed the transfer function of a low-pass filter with a gain A and a pole at an angular frequency ω_p. The gain of the filter is:

$$A = \frac{\alpha_1}{\alpha_2} = \frac{4C_{in}}{C_{FB}} \qquad (6.25)$$

Its cutoff angular frequency is:

$$\omega_p = \frac{\alpha_2}{T_{mod}} = \alpha_2 f_{mod} \qquad (6.26)$$

This corresponds to a pole frequency f_p at:

$$f_p = \frac{\omega_p}{2\pi} = \frac{\alpha_2 f_{mod}}{2\pi} = \frac{C_{FB}}{8\pi C_{sum}} f_{mod} \qquad (6.27)$$

The gain and the pole can thus be set by appropriately choosing the α_1 and α_2 capacitor ratios. In the implementation of the current measurement microsystem, the circuit can be configured with different values of α_1 and α_2 for test purposes (see section 7.1).

4.4 Offset demodulator

The offset can be extracted using the same demodulator circuit as for the external signal (figure 146), using the timing presented in figure 155. It corresponds to the ++++ sequence necessary to extract $4V_{off}$ from the input V_{mod} (see table 16).

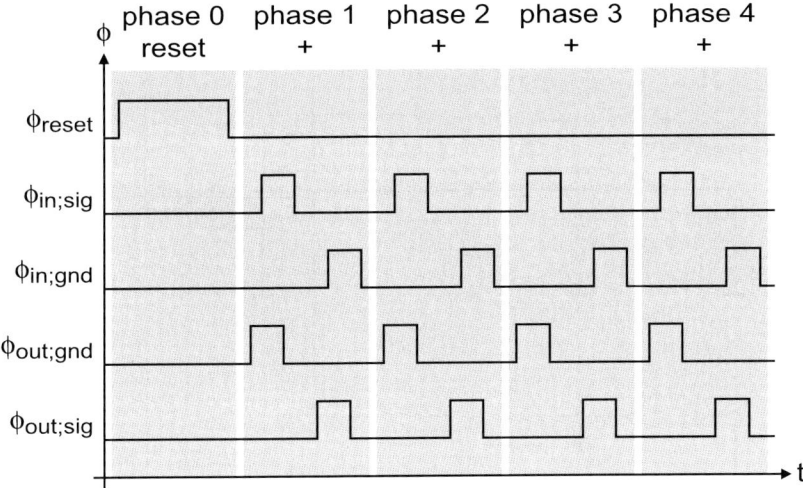

Figure 155. Offset signal demodulator switch timing

As for the external signal extraction, the reset can be included in one of the demodulation phases (see section 4.2).

Since the offset signal is the mean value (DC component) of the modulated output of the preamplifier V_{mod} (chapter 5, section 6.3), the demodulator for the offset component can be simplified by implementing it as a simple analog low-pass filter. If only large offset values need to be detected, it is also possible to just use a comparator to detect situations where the modulated signal is always positive or negative. An extreme simplification is obtained by using a digital inverter as comparator.

Finally, if the offset must not be compensated continuously and its demodulation can be performed alternately with gain calibration, it is possible to share the reference demodulator between offset and gain detection. Another solution is to multiplex the operational amplifier and use it in both reference and offset demodulator, using the technique presented in [103].

5 DELTA-SIGMA MODULATOR

The delta-sigma modulator of chapter 5, section 6.7 is implemented by the switched-capacitor integrator presented in figure 156.

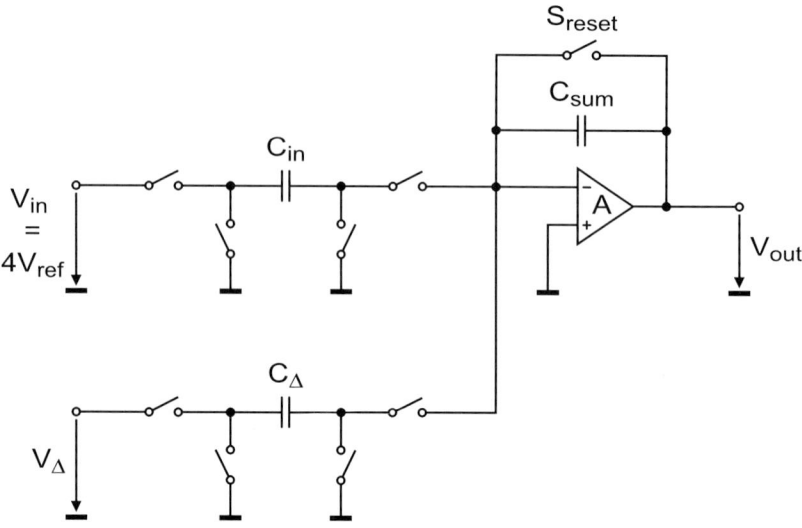

Figure 156. Delta-sigma modulator

The input signal of the delta-sigma modulator is the $4V_{ref}$ demodulated and already low-pass filtered signal output from the reference demodulator. It is sampled once per cycle of 4 demodulation phases. The feedback is performed indirectly by using a digital feedback loop, which sets the sign of the difference signal ($\pm V_\Delta$) to the opposite of the sign of the output V_{out}. The decision is taken at the beginning of each phase, using a comparator connected to the output of the modulator (see also chapter 5, section 6.7). The sign is changed by appropriately operating the 4 switches connected to the C_Δ capacitor.

Figure 157 presents the switch timing of the delta-sigma modulator. For the difference signal (V_Δ) sign, both alternatives are displayed. During operation, the combination of + and - phases is decided dynamically and the 4 corresponding signals are generated for each phase according to the figure.

As for the reference demodulator (see section 4.3), the reset switch is not used, but can be activated one single time at the beginning of each gain calibration cycle.

Chapter 6: Implementation of the Hall microsystem

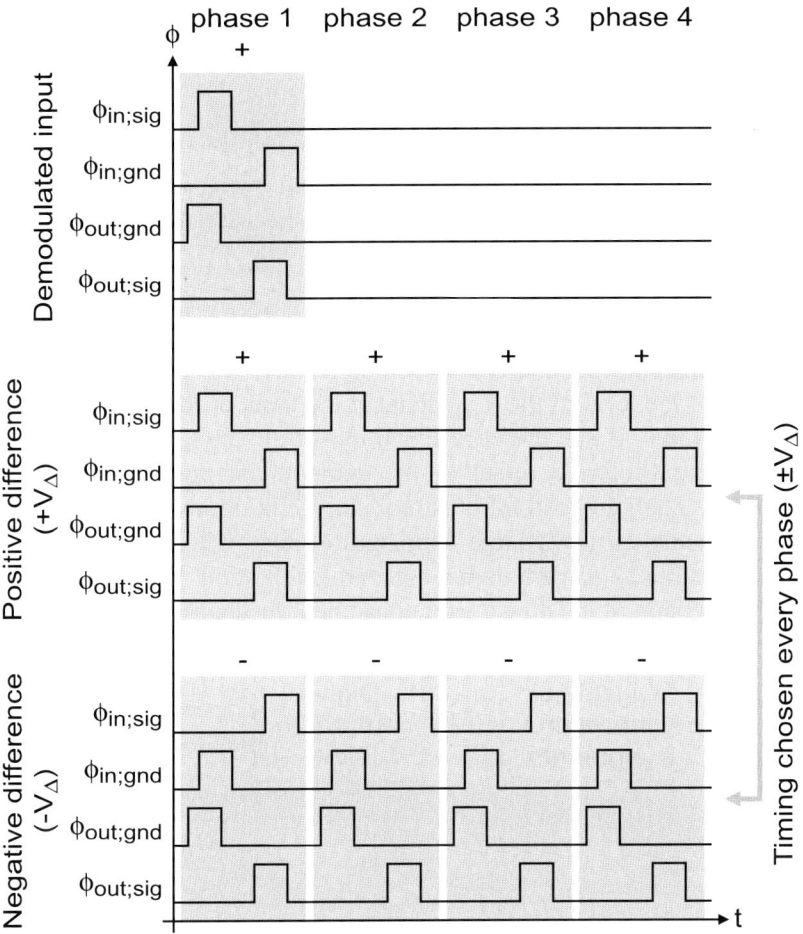

Figure 157. Delta-sigma switch timing

The input signal of the delta-sigma modulator is the $4V_{ref}$ signal produced every 4 demodulation phases by the reference demodulator. The delta-sigma must thus by synchronized with the reference demodulator and sample its output once after 4 demodulation phases. This is achieved if the delta-sigma modulator and the demodulator are synchronized according to figures 154 and 157, i.e. with identical phase numbering.

On the other hand, the difference signal (V_Δ) is added or subtracted 4 times per cycle, once during each phase. An up/down counter is used to compute the difference between the number of positive and negative steps. The value obtained is the digital low-pass filtered gain measurement result, which is compared to the digital nominal value and determines the sign of the compensation according to the principle described in chapter 5, section 6.4.

6 SYSTEM IMPROVEMENTS

The performances of the current measurement microsystem are limited by several block-level imperfections. This section discusses the main issues and their solutions.

6.1 Compensation of the reference demodulator offset

In the gain compensation feedback loop (figure 128) presented in chapter 5, section 6.4, the level of the V_{ref} signal at the input of the demodulator is in the order of 4 mV. If the reference signal must be extracted with a 0.1 % precision, the input-referred offset of the demodulator and delta-sigma (due principally to charge injection) must not change by more than 4 μV. This is difficult to achieve over a large temperature range without calibration, but becomes possible by using a digital autozero calibration. Figure 158 presents a modified version of the digital gain adjustment feedback loop of figure 128.

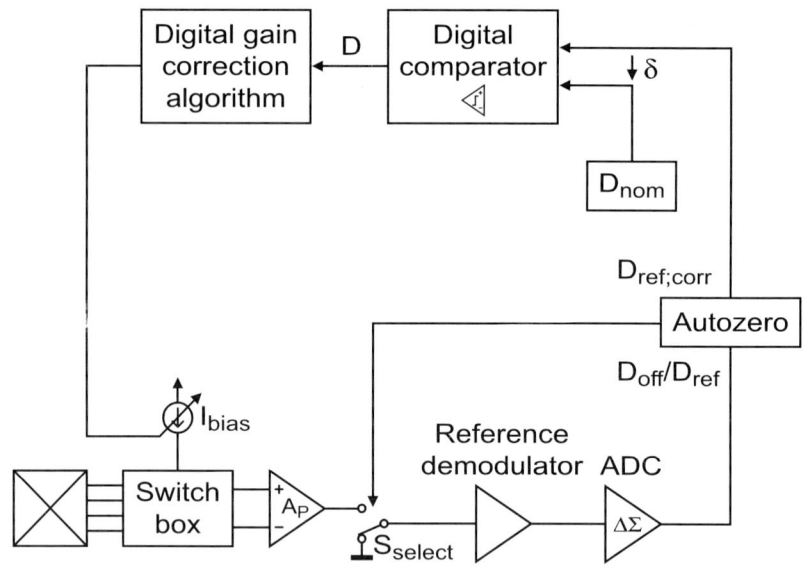

Figure 158. Offset compensation in the gain adjustment feedback loop

Instead of continuously extracting the reference signal $4V_{ref}$ from the modulated signal V_{mod}, the reference demodulator and delta-sigma converter

Chapter 6: Implementation of the Hall microsystem

are also used to measure and calibrate their own offset before the sensitivity measurement. This is achieved by carrying out two successive analog-to-digital conversion phases.

During the first phase, the input of the reference demodulator is connected to ground (as in figure 158). The demodulator and delta-sigma are operated exactly as during the sensitivity measurement, and the result obtained is the digital measurement D_{off} of the offset of the combined demodulator and delta-sigma system.

During the second phase, the input of the reference demodulator is connected to the modulated output V_{mod} of the preamplifier, by toggling the S_{select} switch. The digital result of the second conversion cycle is the gain measurement D_{ref}.

Instead of comparing directly D_{ref} to the nominal value D_{nom}, the corrected value $D_{ref;corr}$ is used for the comparison:

$$D_{ref;corr} = D_{ref} - D_{off} \qquad (6.28)$$

If the offset has the same value during both measurements, the corrected value is offset-free. In fact, the demodulator and delta-sigma are autozeroed: The offset and low-frequency noise are cancelled, whereas an additional foldover component is introduced (see chapter 2, section 4).

6.2 Coil-sensor capacitive coupling

In another microsystem [80], it is reported that the coil-sensor capacitive coupling limits the modulation frequency to about 1 kHz [85]. Figure 159 presents a simplified model of this phenomenon. The coupling is represented only for one of the two sensor terminals (S_-).

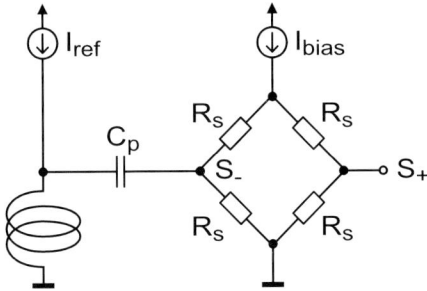

Figure 159. Model of the coil-sensor capacitive coupling

The capacitor C_p represents the direct capacitive parasitic coupling between the coil and the S_ terminal of the Hall sensor. Its value is difficult to estimate, but knowing the dimensions of the coil and sensor, it is certainly smaller than 1 pF. On the other hand, the impedance $R_s/2$ of the sensor terminal is in the order of 500 Ω. The parasitic transfer function is thus high-pass with a zero at a frequency f_z:

$$f_z = \frac{1}{\pi R_s C_p} \geq 630 \text{ [MHz]} \quad (6.29)$$

In [80], no estimation of f_z is given. However, measurement results of the capacitive coupling are presented up to 10 kHz. The zero frequency is not reached, since all the measurements are in the +20 dB/decade region. Despite this fact, frequencies higher than 1 kHz cannot be used in [85], as they generate an excessive parasitic voltage at the sensor terminals.

This limitation does not exist if the modulation technique presented in chapter 5, section 6 is used. In fact, since the coil modulation is a square signal (the polarity of the current is reversed periodically), the important parameter here is the settling time of the Hall sensor voltage. Even if the transient perturbation is high, it is no issue as long as its damping is fast. This is obviously the case since f_z is much higher than the modulation frequency, which is in the order of 1 MHz.

At any rate, it is preferable to use simple design precautions like fixing the internal voltage of the coil [80] to limit the transient amplitude of the spikes.

6.3   External interferences

Current measurement microsystems are often used in perturbed environments. In particular, they are submitted to the fast variations of the external electrical field (dV/dt) near the primary current conductor, which can be in the order of 10 kV/μs [104]. These variations are directly coupled to the internal nodes of the ASIC through parasitic capacitances.

No special design effort has been made in the current measurement microsystem described here to limit the effect of these perturbations. However, when integrating the system into a commercial application, it will be necessary to address this problem.

If a fully-differential circuit is designed, the immunity of the system to external interferences can be improved. If the dual signal lines are located close to each other in the entire circuit, the influence of the external interferences is limited. Since the effect of the interferences is similar on both signals in this case, their difference is indeed almost null.

The folded-cascode amplifiers presented in sections 3.2 and 3.3 can be designed to be fully-differential [105][106][107]. Special care must also be taken in the layout to keep both signal lines as close as possible to each other. This rule must also be applied to the dual components connected to these lines. Differential switched-capacitor versions of the circuits presented in this chapter can be realized [108].

Finally, it is possible to eliminate short transient parasitics by detecting and eliminating them, in particular in the sensitivity measurement loop. In fact, it is not problematic if one erroneous sample (caused by an important dV/dt at a given instant for instance) is sorted out and not fed into the demodulator. This is possible because the reference integration time is long and the signal rate high, which allows eliminating some samples without any problem.

6.4 Alternate modulation/demodulation schemes

Two alternate modulation and demodulation schemes are proposed in this section. The first one is a variant of the basic scheme presented in chapter 5, sections 6.2 and 6.3. It allows to double the frequency of the zero of the parasitic transfer function (see chapter 5, section 6.8). The second proposed scheme is fundamentally different and helps to perform precise sensitivity measurements even with higher external fields.

Table 22 presents a modulation scheme where the reference modulation frequency is twice the spinning frequency, which is exactly the reverse of the basic scheme of chapter 5, section 6.2. Table 23 presents the associated demodulation schemes for the external, reference and offset signals.

The advantage of this alternate scheme is that the zero of the parasitic transfer function is pushed at twice the frequency of the basic scheme. The same development can be done as in chapter 5, section 6.8 with the new scheme. The modulation frequency of V_{ext} is here twice the value of the basic scheme. This is because the modulation and demodulation changes the sign of V_{ext} at every demodulation phase, whereas it is only at every second phase for the basic scheme.

Table 22. Reverse modulation scheme

Phase (i)	Modulation		Preamplifier output ($V_{mod;i}$)
	Spinning	Reference	
1	+	+	$+ (V_{ext} + V_{ref}) + V_{off}$
2	+	−	$+ (V_{ext} - V_{ref}) + V_{off}$
3	−	+	$- (V_{ext} + V_{ref}) + V_{off}$
4	−	−	$- (V_{ext} - V_{ref}) + V_{off}$

Table 23. Reverse demodulation schemes

Phase (i)	Preamplifier output ($V_{mod;i}$)	Demodulation		
		Signal	Reference	Offset
1	$+ (V_{ext} + V_{ref}) + V_{off}$	+	+	+
2	$+ (V_{ext} - V_{ref}) + V_{off}$	+	−	+
3	$- (V_{ext} + V_{ref}) + V_{off}$	−	−	+
4	$- (V_{ext} - V_{ref}) + V_{off}$	−	+	+
		$4 V_{ext}$	$4 V_{ref}$	$4 V_{off}$

When a variation of V_{ext} is considered, the reverse demodulation gives:

$$V_{mod;1} - V_{mod;2} - V_{mod;3} + V_{mod;4} = 4V_{ref} + V_{ext;1} - V_{ext;2} + V_{ext;3} - V_{ext;4} \tag{6.30}$$

For the basic demodulation scheme, the result is:

$$V_{mod;1} - V_{mod;2} - V_{mod;3} + V_{mod;4} = 4V_{ref} + V_{ext;1} + V_{ext;2} - V_{ext;3} - V_{ext;4} \tag{6.31}$$

With the alternate scheme, the modulation is indeed at twice as high frequency and equation 5.81 becomes:

Chapter 6: Implementation of the Hall microsystem

$$A_{LF} = \left\| \frac{d}{dt} e^{j\omega t} \right\| T_{mod} = \omega T_{mod} \qquad (6.32)$$

and the new zero frequency:

$$f_z = \frac{f_{mod}}{2\pi} \qquad (6.33)$$

It is thus advantageous to use this reverse scheme in order to push the zero to higher frequency and thus increase the rejection ratio of the interferences.

Another alternate scheme takes advantage of a sensor array to cancel the external magnetic field at the input of the preamplifier (see section 2). It can be used as a complement to the filtering of the interferences due to the external field, or as an alternative if the external field is very high (e.g. 500 mT instead of 50 mT). Table 24 presents a multiplexed modulation scheme where the system is used alternately for external signal and reference measurement.

Table 24. Multiplexed modulation scheme

Phase (i)	Modulation		Preamplifier output ($V_{mod;i}$)
	Spinning	Reference	
1	+	0	$V_{ext} + V_{off}$
2	−	0	$-V_{ext} + V_{off}$
3	+	+	$V_{ref} + V_{off}$
4	−	+	$-V_{ref} + V_{off}$

During phases 1 and 2, the external magnetic field is measured alone (the reference is off) and the spinning current technique allows offset cancellation. During phases 3 and 4, the sensor array cancels the external magnetic field (section 2) and the reference signal is amplified. The signal, reference and offset components can be extracted using the demodulation schemes described in table 25.

Table 25. Multiplexed demodulation scheme

Phase (i)	Preamplifier output ($V_{mod;i}$)	Demodulation		
		Signal	Reference	Offset
1	$V_{ext} + V_{off}$	+	0	+
2	$-V_{ext} + V_{off}$	-	0	+
3	$V_{ref} + V_{off}$	0	+	+
4	$-V_{ref} + V_{off}$	0	-	+
		$2 V_{ext}$	$2 V_{ref}$	$4 V_{off}$

The external signal is extracted during the phases 1 and 2, whereas the reference is measured only during phases 3 and 4 with a higher gain. The system is thus multiplexed and its principle becomes similar to the twin-sensor circuit of chapter 5, section 4.4. The difference, however, is that the sensor used for measurement is still the one that is calibrated, whereas in [85] a second matched system is necessary and indirect sensitivity measurement is performed.

7 SYSTEM INTEGRATION

The system has been integrated to validate the continuous sensitivity calibration concept. The circuit has been designed for testability [109], i.e. to ensure maximum flexibility for configuration and measurement possibilities. This choice limits the performances of the system, mainly in terms of power consumption which is higher than strictly necessary due to additional test circuitry.

7.1 Configuration and measurement possibilities

The configuration capabilities of the circuit can be separated in two categories. Firstly, circuit parameters of some internal blocks can be modified by component selection and timing variations. In the reference demodulator, for instance, the gain value and pole frequency can be modified thanks to a set of capacitors or by changing the switched-capacitor control signals. Secondly, internal signals in critical nodes can be output and observed outside the circuit, and external signals fed into internal blocks. This allows to thoroughly test the circuit in a variety of conditions and to analyze the performances of separate blocks.

Table 26 shows the component values that can be combined in the reference demodulator of figure 153. The elementary values can be added up by using the capacitors in parallel if desired.

Table 26. Capacitor values in the reference demodulator

Capacitor	Value [pF]
C_{in}	2, 5
C_{FB}	0.05, 0.2, 0.5
C_{Cref}	0.2, 1
C_{sum}	20

Using these values in equations 6.25 and 6.27, the minimum pole frequency is $f_{p;min} \cong 100$ Hz ($C_{FB} = 0.05$ pF). The corresponding maximum gain is $A_{max} = 400$ ($C_{in} = 5$ pF). Other combinations of components give higher pole frequencies and lower gains.

For all the switched-capacitor circuits in the microsystem, it is possible to generate specific timings, since each switch control signal is accessible separately on a dedicated pad. This provides maximum flexibility for testing different modulation and demodulation schemes, and observing their influence on system performance.

The second category of system configuration capabilities is implemented by analog switches, which route signals inside the circuit. In the preamplifier, three different gains are available: 10, 100 and 1000. For each one of the 3 parallel demodulators, the gain can be selected independently. The output of the demodulators and delta-sigma can be fed outside the circuit through a buffer. Finally, it is also possible to use the demodulators with an external modulated signal instead of the preamplified Hall sensor voltage.

7.2 Integrated circuit

The circuit has been integrated in a conventional 0.8 μm 5V CMOS technology, and encapsulated in a standard 84 pins CLCC package. Figure 160 presents a micrograph of the current measurement microsystem.

Figure 160. Micrograph of the current measurement microsystem

The complete circuit including the padring occupies 11.5 mm². Table 27 summarizes the classification of the pin attributions.

Separate analog and digital power supplies are used to improve the immunity of the analog circuits to noise. Slightly more than 25 % of the pads are analog, the remainder being digital.

Chapter 6: Implementation of the Hall microsystem   233

Table 27. Pin functions

	Description	Number of pins
Analog	Power supply	3
	Coils	4
	Bias currents	3
	Compensation currents	4
	Reference voltages	3
	External signal inputs	2
	Buffered outputs	4
Digital	Power supply	2
	Spinning current	4
	Configuration inputs	15
	Switched-capacitor switch controls	39
	Delta-sigma output	1
	Total	84

In the analog part, the offset of the DDA (and indirectly, the offset of the sensor) can be compensated, as well as the offset of the demodulators and the delta-sigma. The 3 demodulators can be observed simultaneously and the delta-sigma has both an analog and a digital output. Finally, the preamplified signal can be measured at every gain stage.

In the digital part, the flexibility is achieved thanks to the 39 parallel switch control signals. There are also configuration inputs for the analog switches routing the signals inside the system.

7.3   Measurement results

The functionality of the blocks composing the system was tested and no problem was detected. The system parameters were then verified, and the principle and the feasibility of the calibration demonstrated. The circuit behaves as outlined in the theoretical description above.

Figure 161 presents the measured preamplifier and external signal demodulator output voltages as a function of time, for the case where no external magnetic field is applied ($B_{ext} = 0$). The preamplifier gain (see section 3.1) is programmed to be 100. The modulated preamplified signal V_{mod} is demodulated by the external signal demodulator (section 4.2) in 4 phases, each one having a duration of 1 µs. Two successive demodulation cycles are presented in the figure. The demodulation scheme presented in table 16 (signal demodulation column) is used to produce the intermediate results $V_{out;1}, ..., V_{out;4}$ of table 21, and the final result $V_{out;4}$ is the demodulated external component $4V_{ext}$ (equation 6.17). The switch timings of figure 151 are used to control the demodulator. However, phase 0 (reset) is merged with phase 1, by performing the reset during the first step of phase 1 (see section 4.2).

The sensor and preamplifier offset is -3 mV, and consequently $V_{off} = -0.3$ V (the preamplifier gain is 40 dB). The reference signal V_{ref} is negligible (4 mV) and cannot be observed in the figure. All voltages are referred to the internal ground of the microsystem, which is equal to 2.5 V (half the power supply voltage) and represented by a gray line in the figure.

Table 28 presents the analytical and numerical values of the output of the demodulator at the end of each phase. Since $V_{ext} = 0$, the final value $V_{out;4}$ is also null.

Figure 162 and table 29 present the case of a negative external magnetic field. The corresponding preamplified voltage V_{ext} is -0.18 V. In figure 163 and table 30, the case of a positive external magnetic field is displayed, with a corresponding preamplified voltage V_{ext} of 0.18 V.

Chapter 6: Implementation of the Hall microsystem

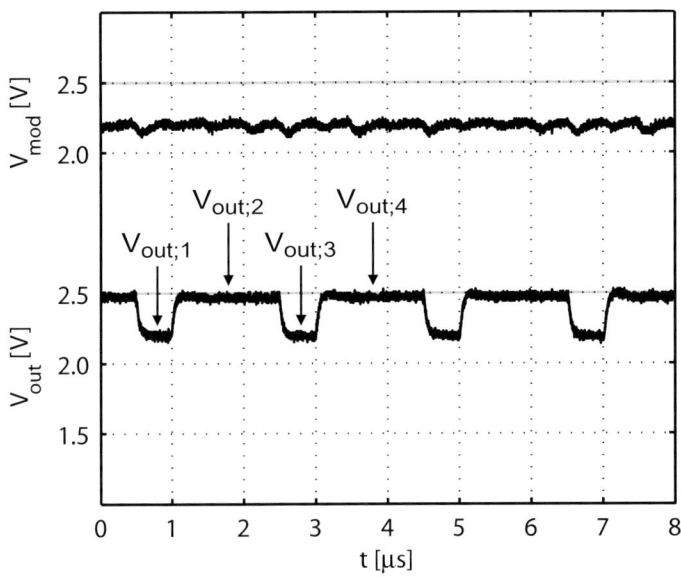

Figure 161. Preamplifier and demodulator output for $B_{ext} = 0$

Table 28. Demodulator output for $B_{ext} = 0$
($V_{ext} = 0$ [V]; $V_{off} = -0.3$ [V]; V_{ref} negligible)

Phase (i)	$V_{out;i}$ (analytical)	$V_{out;i}$ (numerical) [V]
1	$V_{off} + V_{ref} \cong V_{off}$	2.5 - 0.3 = 2.2
2	$0 + 2V_{ref} \cong 0$	2.5 + 0.0 = 2.5
3	$V_{off} + V_{ref} \cong V_{off}$	2.5 - 0.3 = 2.2
4	0	2.5 + 0.0 = 2.5

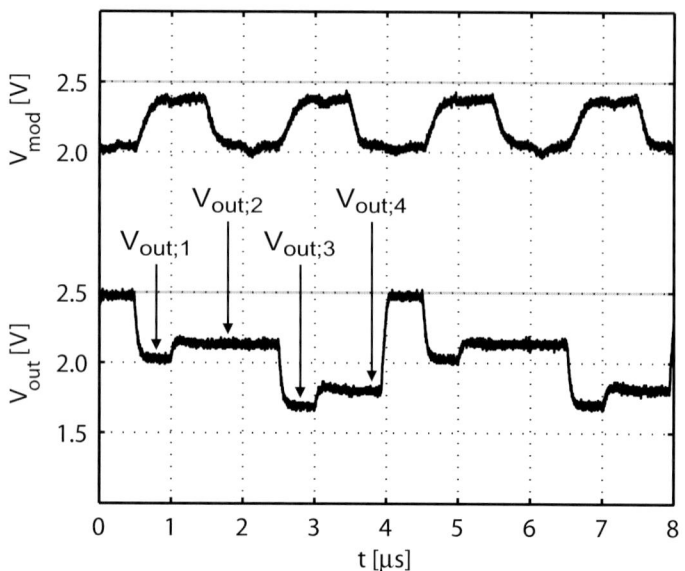

Figure 162. Preamplifier and demodulator output for negative B_{ext}

Table 29. Demodulator output for negative B_{ext}
(V_{ext} = -0.18 [V]; V_{off} = -0.3 [V]; V_{ref} neglected)

Phase (i)	$V_{out;i}$ (analytical)	$V_{out;i}$ (numerical) [V]
1	$V_{ext} + V_{off}$	2.5 - 0.18 - 0.3 = 2.02
2	$2V_{ext}$	2.5 - 0.36 = 2.14
3	$3V_{ext} + V_{off}$	2.5 - 0.54 - 0.3 = 1.66
4	$4V_{ext}$	2.5 - 0.72 = 1.78

Chapter 6: Implementation of the Hall microsystem

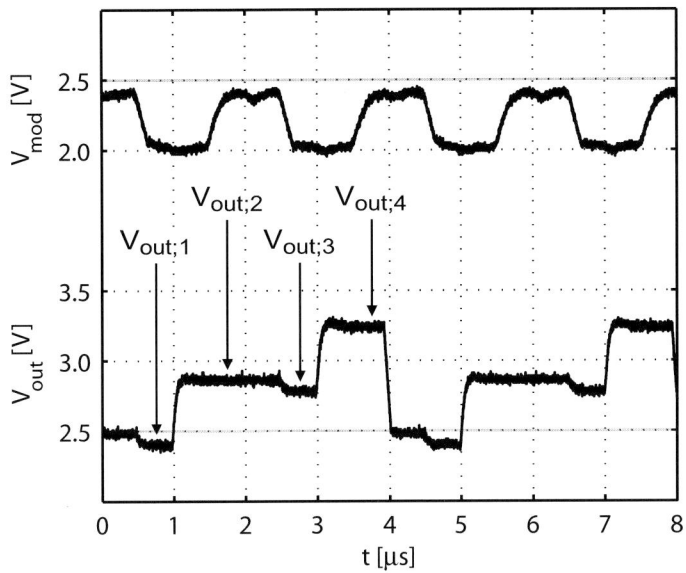

Figure 163. Preamplifier and demodulator output for positive B_{ext}

Table 30. Demodulator output for positive B_{ext}
(V_{ext} = 0.18 [V]; V_{off} = -0.3 [V]; V_{ref} neglected)

Phase (i)	$V_{out;i}$ (analytical)	$V_{out;i}$ (numerical) [V]
1	$V_{ext} + V_{off}$	2.5 + 0.18 - 0.3 = 2.38
2	$2V_{ext}$	2.5 + 0.36 = 2.86
3	$3V_{ext} + V_{off}$	2.5 + 0.54 - 0.3 = 2.74
4	$4V_{ext}$	2.5 + 0.72 = 3.22

The input/output characteristics of the microsystem were studied in detail. In particular, the nonlinearity over the complete measurement range of ± 50 mT was measured on one sample at room temperature using a fully automatic measurement setup. Figure 164 presents the results obtained, with a worst-case nonlinearity lower than 0.025 %.

Figure 164. Nonlinearity measurement

The offset drift of 3 samples as a function of temperature was measured at temperatures ranging from -40 °C to 80 °C, using a fully automatic measurement setup including a programmable climatic chamber. The results are presented in figure 165. The achieved offset drift is lower than 50 µV/°C.

Finally, with the same measurement setup, the efficiency of the sensitivity calibration was verified by measuring the gain drift of 3 microsystems as a function of temperature. Figure 166 presents the results obtained.

A typical sensitivity drift of 30 ppm/°C is achieved over the full temperature range, and the worst-case drift is 50 ppm/°C. Table 31 summarizes the most important characteristics of the microsystem.

The sensitivity drift is comparable to the result obtained in [85]. Another feature is the high bandwidth. If an analog-to-digital converter is connected directly to the output of the signal demodulator without low-pass filtering, it can be as high as 500 kHz. The remaining characteristics are similar to existing current measurement microsystems (see table 14).

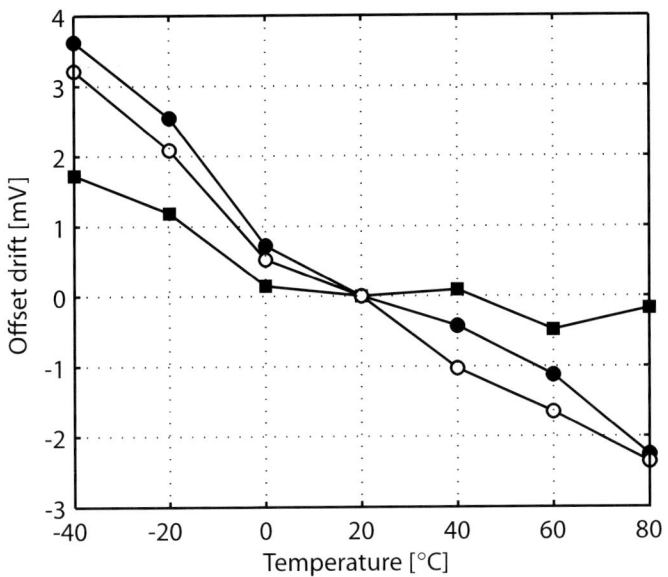

Figure 165. Offset drift measurement

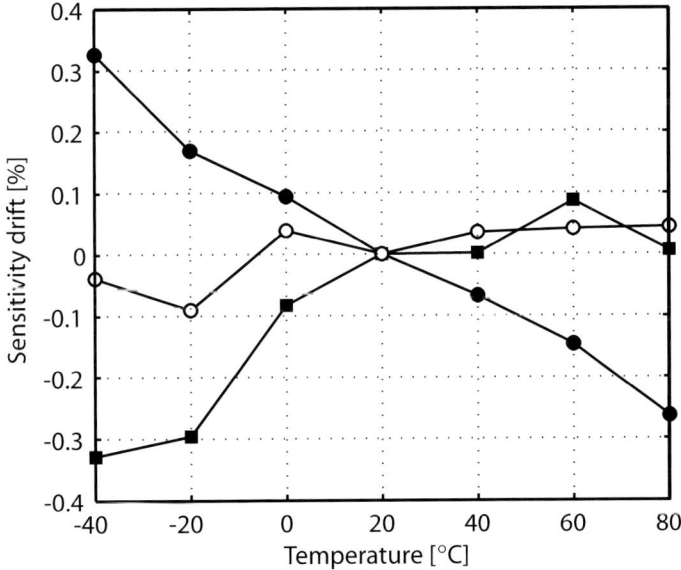

Figure 166. Sensitivity drift measurement

Table 31. Microsystem characteristics

Parameter	Value	Unit
Supply voltage	5	V
Sensitivity	35	V/T
Measurement range	± 50	mT
Bandwidth	500	kHz
Nonlinearity	< 0.1	%
Offset drift	< 50	µV/°C
Gain drift	< 50	ppm/°C

8   CONCLUSION

The system implementation uses all the digital calibration techniques presented in the previous chapters. The test circuit proves the feasibility of the continuous sensitivity calibration technique. A low residual thermal drift of 50 ppm/°C is achieved using the proposed correction principle. This is 6 to 10 times less than drifts currently achieved in commercial Hall sensor analog front-ends.

Chapter 7

Conclusion

This book has two main parts: a theoretical study and practical realizations. The achievement of the theoretical study is a complete digital compensation methodology. This methodology is successfully used in three totally different practical implementations.

To summarize and conclude this book, the highlights, main contributions and further development perspectives are presented and discussed.

1 HIGHLIGHTS

In the first part of this book, the state of the art and a systematic analysis of the existing compensation techniques are presented. The chopper and autozero techniques are compared and their advantages and drawbacks discussed. In particular, their compatibility with sampled and continuous-time analog circuits is examined and classificated.

Then, a complete review of sub-binary converters is proposed. The $M/2^+M$ structure is thoroughly analyzed, from circuit theory to design issues, including layout. The different structures of sub-binary converters are compared.

A **complete digital compensation methodology** based on very low-area sub-binary $M/2^+M$ current-mode converters and successive approximations algorithms is then proposed. The systematic approach describes the compensation of circuit imperfections from their detection to their compensation. A simulation tool is introduced, allowing the automatic and transparent simulation of analog circuits that include digital compensation blocks. By using two $M/2^+M$ and a special current mirror, an **up/down DAC** for digital compensation of continuous-time systems is also presented.

Three applications of the compensation methodology and circuits are then proposed. Two additional calibration and radix conversion algorithms allow the use of a $M/2^+M$ converter as a **high-precision DAC**. The complete converter fits in a very low area.

The second application is a **SOI 1T DRAM**, where the compensation technique allows to generate a reference current for reading the memory which compensates several circuit imperfections together. Among others, it cancels the effect of the sense amplifier offset, the statistical distribution of the memory cell read currents and other device mismatches.

The main application of the digital compensation application is a **Hall microsystem for current measurement**. In this system, the sensitivity is digitally calibrated using the techniques detailed in the first part: chopper- and autozero-like techniques, an up/down DAC and a new combined modulation technique. The calibration cancels the sensitivity variations due to temperature, mechanical stresses and ageing. In particular, a sensitivity thermal drift of less than 50 ppm/°C is achieved. This is 6 to 10 times less than in current commercial implementations.

2 MAIN CONTRIBUTIONS

The main contributions of this book are:

- The systematic analysis of the $M/2^+M$ structure and its use in conjunction with a successive approximations algorithm for the digital compensation of analog circuit imperfections.
- The automatic 2-pass simulation technique allowing the transparent simulation of analog circuits including a digital compensation.
- The use of a $M/2^+M$ structure as a conventional DAC, using two calibration and radix conversion algorithms. This is an adaptation of a well-known technique used in pipelined converters.
- The up/down DAC for digital compensation in continuous-time systems.
- The automatic reference calibration in a DRAM, and in particular in a SOI 1T DRAM. Usually, such memory circuits are not calibrated.
- The sensitivity measurement and compensation technique in Hall sensor-based microsystems, in particular the combined modulation and demodulation schemes.

3 PERSPECTIVES

Although quite some work has been done, there is still a lot to do. A few ideas are briefly discussed below.

First, the compensation methodology can be integrated in a new CAD tool, or into an extension of an existing one. Besides helping the designer with automatic and transparent simulation, it also guides him to make tradeoffs.

Chapter 7: Conclusion

The DAC based on a M/2$^+$M converter is a side discovery of this work. With more time, this circuit must be analyzed in detail and its tradeoffs, limitations and possible optimizations studied.

The SOI 1T DRAM is still the object of intense work towards the realization of an industrial product. A world patent already protects the calibration principle and a team of designers works on circuit implementations.

Finally, the sensitivity calibration in Hall sensor microsystems will also be used in future commercial product, but there are still many points to be studied before. Firstly, by analysing the current prototype in more detail, the performances of the system may still be improved. Secondly, a precise and driftless integrated reference for the sensitivity must be developed. Finally, the other external features, including the digital signal generators and bias sources, must also be integrated to obtain a final product with a few pins only: 2 for the power supply and 1 for the measurement output, ideally.

References

[1] E. A. Vittoz, "The Design of High-Performance Analog Circuits on Digital CMOS Chips", IEEE Journal of Solid-State Circuits, Vol. 20, pp. 657-665, June 1985

[2] J-B. Shyu, G. C. Temes, F. Krummenacher, "Random Error Effects in Matched MOS Capacitors and Current Sources", IEEE Journal of Solid-State Circuits, Vol. 19, pp. 948-955, December 1984

[3] M. J. M. Pelgrom, C. J. Duinmaijer, A. P. G. Welbers, "Matching Properties of MOS Transistors", IEEE Journal of Solid-State Circuits, Vol. 24, pp. 1433-1440, October 1989

[4] C. Michael, M. Ismail, "Statistical Modeling of Device Mismatch for Analog MOS Integrated Circuits", IEEE Journal of Solid-State Circuits, Vol. 27, pp. 154-166, February 1992

[5] C. C. Enz, E. A. Vittoz, F. Krummenacher, "A CMOS Chopper Amplifier", IEEE Journal of Solid-State Circuits, Vol. 22, pp. 335-342, June 1987

[6] C. C. Enz, G. C. Temes, "Circuit Techniques for Reducing the Effects of Op-Amp Imperfections: Autozeroing, Correlated Double Sampling, and Chopper Stabilization", Proceedings of the IEEE, Vol. 84, pp. 1584-1614, November 1996

[7] P. E. Allen, D. R. Holberg, "CMOS Analog Circuit Design", Second edition, Oxford University Press, ISBN 0-19-511644-5, 2002

[8] G. Wegmann, E. A. Vittoz, F. Rahali, "Charge Injection in Analog MOS Switches", IEEE Journal of Solid-State Circuits, Vol. 22, pp. 1091-1097, December 1987

[9] K. Martin, A. S. Sedra, "Effects of the Op Amp Finite Gain and Bandwidth on the Performance of Switched-Capacitor Filters", IEEE Transactions on Circuits and Systems, Vol. 28, pp 822-829, August 1981

[10] M. H. White, D. R. Lampe, F. C. Blaha, I. A. Mack, "Characterization of Surface Channel CCD Image Arrays at Low Light Levels", IEEE Journal of Solid-State Circuits, Vol. 9, pp. 1-13, February 1974

[11] I. Opris, G. Kovacs, "A Rail-to-Rail Ping-Pong Op-Amp", IEEE Journal of Solid-State Circuits, Vol. 31, pp. 1320-1324, September 1996

[12] M. Kayal, R. T. L. Sáez, M. Declercq, "An automatic Offset Compensation Technique Applicable to Existing Operational Amplifier Core Cell", IEEE Custom Integrated Circuits Conference, pp 419-422, May 1998

[13] C.-G. Yu, R. L. Geiger, "An Automatic Offset Compensation Scheme with Ping-Pong Control for CMOS Operational Amplifiers", IEEE Journal of Solid-State Circuits, Vol 29, pp. 601-610, May 1994

[14] Y. Huang, G. C. Temes, P. F. Ferguson, "Novel high-frequency track-and-hold stages with offset and gain compensation", IEEE International Symposium on Circuits and Systems, Vol. 1, pp. 155-158, May 1996

[15] R. V. D. Plassche, "Dynamic Element Matching for High-Accuracy Monolithic D/A Converters", IEEE Journal of Solid-State Circuits, Vol. 11, pp. 795-800, December 1976

[16] U. K. Moon, G. C. Temes, J. Steensgaard, "Digital Techniques for Improving the Accuracy of Data Converters", IEEE Communications Magazine, Vol. 37, pp. 136-143, October 1999

[17] A. T. K. Tang, "A 3µV-Offset Operational Amplifier with 20 nV/√Hz Input Noise PSD at DC Employing both Chopping and Autozeroing", IEEE International Solid-State Circuits Conference, Vol. 1, pp. 386-387, February 2002

[18] P. Hasler, B. A. Minch, C. Diorio, "Floating-gate devices: they are not just for digital memories anymore", IEEE International Symposium on Circuits and Systems, Vol. 2, pp. 388-391, May 1999

[19] L. R. Carley, "Trimming Analog Circuits Using Floating-Gate Analog MOS Memory", IEEE Journal of Solid-State Circuits, Vol. 24, pp. 1569-1575

[20] J. E. Franca, F. Nunes, "Successive approximation tuning of monolithic continuous-time filters", Electronics Letters, Vol. 28, pp. 1696-1697, August 1992

[21] J. Goes, J. Franca, N. Paulino, "High-Linearity Calibration of Low-Resolution Digital-to-Analog Converters", IEEE International Symposium on Circuits and Systems, Vol. 5, pp. 345-348, May 1994

[22] C. A. Leme, J. E. Franca, "Efficient calibration of binary weighted networks using a mixed analogue-digital RAM", IEEE International Symposium on Circuits and Systems, Vol. 3, pp.1545-1548, June 1991

[23] W. G. Bliss, C. E. Seaberg, R. L. Geiger, "A Very Small Sub-Binary Radix DAC for Static Pseudo-Analog High-Precision Memory", Proceedings of the 35th Midwest Symposium on Circuits and Systems, Vol. 1, pp. 425-428, August 1992

[24] M. P. Kennedy, "On the Robustness of R-2R Ladder DACs", IEEE Transactions on Circuits and Systems, Part I: Fundamental Theory and Applications, Vol. 47, pp. 109-116, February 2000

References

[25] T. S. Rathore, A. Jain, "Abundance of Ladder Digital-to-Analog Converters, IEEE Transactions on Instrumentation and Measurement, Vol. 50, pp. 1445-1449, October 2001

[26] K. Bult, G. J. M. G. Geelen, "An Inherently Linear and Compact MOST-Only Current Division Technique", IEEE Journal of Solid-State Circuits, Vol. 27, pp. 1730-1735, December 1992

[27] E. A. Vittoz, X. Arreguit, "Linear networks based on transistors", Electronics Letters, Vol. 29, pp. 297-299, February 1993

[28] M. Bucher, "Analytical MOS transistor modelling for analog circuit simulation", PhD thesis N° 2114, Ecole Polytechnique Fédérale de Lausanne (EPFL), Lausanne, Switzerland, 2000

[29] C. Hammerschmied, Q. Huang, "Design and Implementation of an Untrimmed MOSFET-Only 10-Bit A/D Converter with -79-dB THD", IEEE Journal of Solid-State Circuits, Vol. 33, pp. 1148-1157, August 1998

[30] B. Linares-Barranco, T. Serrano-Gotarredona, R. Serrano-Gotarredona, G. Vicente-Sánchez, "On Mismatch Properties of MOS and Resistors Calibrated Ladder Structures", IEEE International Symposium on Circuits and Systems, Vol. 1, pp. 377-380, May 2004

[31] T. B. Tarim, M. Ismail, "Application of a Statistical Design Methodology to Low Voltage Analog MOS Integrated circuits", IEEE International Symposium on Circuits and Systems, Vol. 4, pp. 117-120, May 2000

[32] L. Wang, Y. Fukatsu, K. Watanabe, "Characterization of Current-Mode CMOS R-2R Ladder Digital-to-Analog Converters", IEEE Transactions on Instrumentation and Measurement, Vol. 50, pp. 1781-1786, December 2001

[33] B. Vargha, I. Zoltán, "Calibration Algorithm for Current-Output R-2R Ladders", IEEE Transactions on Instrumentation and Measurement, Vol. 50, pp. 1216-1220, October 2001

[34] A. Biman, D. G. Nairn, "Trimming of Current Mode DACs by Adjusting V_t", IEEE International Symposium on Circuits and Systems, Vol. 1, pp.33-36, May 1996

[35] C. H. J. Mensink, B. Nauta, "CMOS tunable linear current divider", Electronics Letters, Vol. 32, pp. 889-890, May 1996

[36] G. Scandurra, C. Ciofi, "R-βR Ladder Networks for the Design of High-Accuracy Static Analog Memories", IEEE Transactions on Circuits and Systems, Part I: Fundamental Theory and Applications, Vol. 50, pp. 605-612, May 2003

[37] M. Pastre, M. Kayal, "High-precision DAC based on a self-calibrated sub-binary radix converter", IEEE International Symposium on Circuits and Systems, Vol. 1, pp. 341-344, May 2004

[38] E. G. Soenen, R. L. Geiger, "A fully digital self-calibration method for high resolution, pipelined A/D converters", IEEE Midwest Symposium on Circuits and Systems, Vol. 1, pp. 228-231, August 1993

[39] A. N. Karanicolas, H. S. Lee, K. L. Bacrania, "A 15-b 1-Msample/s Digitally Self-Calibrated Pipeline ADC", IEEE Journal of Solid-State Circuits, Vol. 28, pp. 1207-1215, December 1993

[40] E. G. Soenen, R. L. Geiger, "An Architecture and An Algorithm for Fully Digital Correction of Monolithic Pipelined ADC's", IEEE Transactions on Circuits and Systems, Part II: Analog and Digital Signal Processing, Vol. 42, pp. 143-153, March 1995

[41] Y. Cong, R. L. Geiger, "A 1.5-V 14-Bit 100-MS/s Self-Calibrated DAC", IEEE Journal of Solid-State Circuits, Vol. 38, pp. 2051-2060, December 2003

[42] Z. G. Boyacigiller, B. Weir, P. D. Bradshaw, "An Error-Correcting 14b/20µs CMOS A/D Converter", IEEE International Solid-State Circuits Conference, pp. 62-63, February 1981

[43] G. Miller, M. Timko, H. S. Lee, E. Nestler, M. Mueck, P. Ferguson, "An 18b 10µs Self-Calibrating ADC", IEEE International Solid-State Circuits Conference, pp. 168-169, 292, February 1990

[44] D. Stefanovic, M. Kayal, M. Pastre, V. B. Litovski, "Procedural Analog Design (PAD) Tool", IEEE International Symposium on Quality Electronic Design, pp. 313-318, March 2003

[45] M. C. W. Coln, "Chopper Stabilization of MOS Operational Amplifiers Using Feed-Forward Techniques", IEEE Journal of Solid-State Circuits, Vol. 16, pp. 745-748, December 1981

[46] I. G. Finvers, J. W. Haslett, F. N. Trofimenkoff, "A High Temperature Precision Amplifier", IEEE Journal of Solid-State Circuits, Vol. 30, pp. 120-128, February 1995

[47] J. H. Atherton, H. T. Simmonds, "An Offset Reduction Technique for Use with CMOS Integrated Comparators and Amplifiers", IEEE Journal of Solid-State Circuits, Vol. 27, pp. 1168-1175, August 1992

[48] C. A. Leme, J. E. Franca, "Analog-Digital Design in Submicrometric Digital CMOS Technologies", IEEE International Symposium on Circuits and Systems, Vol. 1, pp. 453-456, June 1997

[49] C. A. Leme, J. Silva, P. Rodrigo, J. E. da Franca, "A Low-Power CMOS Nine-Channel 40-MHz Binary Detection System with Self-Calibrated 500-µV Offset", IEEE Journal of Solid-State Circuits, Vol. 33, pp. 565-572, April 1998

[50] M. Degrauwe, E. Vittoz, I. Verbauwhede, "A Micropower CMOS-Instrumentation Amplifier", IEEE Journal of Solid-State Circuits, Vol. 20, pp. 805-807, June 1985

[51] M. Kayal, Z. Randjelovic, "Auto-zero differential difference amplifier", Electronics Letters, Vol. 36, pp. 695-696, April 2000

[52] M. Kayal, E. Chevallaz, F. Burger, R. Popovic, "Magnetic Angular Encoder Using an Automatic Offset Compensation Technique", IEEE Sensors, vol. 2, pp. 767-770, June 2002

References

[53] F. Krummenacher, R. Vafadar, A. Ganesan, V. Valence, "A High-Performance Autozeroed CMOS Opamp with 50μV Offset", IEEE International Solid-State Circuits Conference, pp. 350-351, February 1997

[54] M. Kayal, M. Pastre, M. Blagojevic, L. Portmann, M. Declercq, "Reference Current Generator, and Method of Programming, Adjusting and/or Operating Same", EPFL (Switzerland), World patent, N° WO2004102631, November 2004

[55] S. Okhonin, M. Nagoga, J. M. Sallese, P. Fazan, "A Capacitor-Less 1T-DRAM Cell", IEEE Electron Device Letters, Vol. 23, pp. 85-87, February 2002

[56] M. Blagojevic, M. Pastre, M. Kayal, P. Fazan, S. Okhonin, M. Nagoga, M. Declercq, "SOI Capacitor-Less 1-Transistor DRAM Sensing Scheme with Automatic Reference Generation", IEEE Symposium on VLSI Circuits, pp. 182-183, June 2004

[57] P. C. Fazan, S. Okhonin, M. Nagoga, J. M. Sallese, "A Simple 1-Transistor Capacitor-Less Memory Cell for High Performance Embedded DRAMs", IEEE Custom Integrated Circuits Conference, pp. 99-102, May 2002

[58] M. Blagojevic, "SOI Mixed Mode Design Techniques and Case Studies", PhD thesis N° 3319, Ecole Polytechnique Fédérale de Lausanne (EPFL), Lausanne, Switzerland, 2005

[59] P. Fazan, S. Okhonin, M. Nagoga, J. M. Sallese, L. Portmann, R. Ferrant, M. Kayal, M. Pastre, M. Blagojevic, A. Borschberg, M. Declercq, "Capacitor-Less 1-Transistor DRAM", IEEE International SOI Conference, pp. 10-13, October 2002

[60] J. E. Lenz, "A Review of Magnetic Sensors", Proceedings of the IEEE, Proceedings of the IEEE, Vol. 78, pp. 973-989, June 1990

[61] H. P. Baltes, R. S. Popovic, "Integrated Semiconductor Magnetic Field Sensors", Proceedings of the IEEE, Vol. 74, pp. 1107-1132, August 1986

[62] J. Heremans, "Solid state magnetic field sensors and applications", Journal of Physics D: Applied Physics, Vol. 26, pp. 1149-1168, August 1993

[63] R. S. Popovic, "Hall Effect Devices", Second edition, Institute of Physics Publishing, ISBN 0-7503-0855-9, 2004

[64] R. S. Popovic, "Sensor Microsystems", IEEE International Conference on Microelectronics, Vol. 2, pp. 531-537, September 1995

[65] R. S. Popovic, "Hall Devices for Magnetic Sensor Microsystems", IEEE International Conference on Solid-State Sensors and Actuators, Vol. 1, pp. 377-380, June 1997

[66] P. C. de Jong, F. R. Riedijk, J. van der Meer, "Smart Silicon Sensors - Examples of Hall-effect Sensors", IEEE Sensors, Vol. 2, pp. 1440-1444, June 2002

[67] E. H. Hall, "On a New Action of the Magnet on Electric Currents", American Journal of Mathematics, Vol. 2, pp.287-292, November 1879

[68] J. B. Kammerer, L. Hebrard, V. Frick, P. Poure, F. Braun, "Horizontal Hall Effect Sensor With High Maximum Absolute Sensitivity", IEEE Sensors Journal, Vol. 3, pp. 700-707, December 2003

[69] Z. Randjelovic, "Low-Power High Sensitivity Integrated Hall Magnetic Sensor Microsystems", Hartung-Gorre, ISBN 3-89649-617-4, 2000

[70] E. Jovanovic, T. Pesic, D. Pantic, "3D Simulation of Cross-Shaped Hall Sensor and Its Equivalent Circuit Model", IEEE International Conference on Microelectronics, Vol. 1, pp. 235-238, May 2004

[71] A. Bilotti, G. Monreal, R. Vig, "Monolithic Magnetic Hall Sensor Using Dynamic Quadrature Offset Cancellation", IEEE Journal of Solid-State Circuits, Vol. 32, pp. 829-836, June 1997

[72] R. Steiner, C. Maier, M. Mayer, S. Bellekom, H. Baltes, "Influence of Mechanical Stress on the Offset Voltage of Hall Devices Operated with Spinning Current Method", IEEE Journal of Microelectromechanical Systems, Vol. 8, pp. 466-472, December 1999

[73] A. Bilotti, G. Monreal, "Chopper-Stabilized Amplifiers with a Track-and-Hold Signal Demodulator", IEEE Transactions on Circuits and Systems, Part I: Fundamental Theory and Applications, Vol. 46, pp. 490-495, April 1999

[74] G. Grandi, M. Landini, "A Magnetic Field Transducer Based on Closed-Loop Operation of Magnetic Sensors", IEEE International Symposium on Industrial Electronics, Vol. 2, pp. 600-605, July 2002

[75] D. Manic, J. Petr, R. S. Popovic, "Short and Long-Term Stability Problems of Hall Plates in Plastic Packages", IEEE International Reliability Physics Symposium, pp. 225-230, April 2000

[76] D. Manic, "Drift in Silicon Integrated Sensors and Circuits due to Thermo-Mechanical Stresses", Hartung-Gorre, ISBN 3-89649-591-7, 2000

[77] J. Trontelj, L. Trontelj, R. Opara, A. Pletersek, "CMOS Integrated Magnetic Field Source Used as a Reference in Magnetic Field Sensors on Common Substrate", IEEE Instrumentation and Measurement Technology Conference, Vol. 2, pp. 461-463, May 1994

[78] J. Trontelj, R. Opara, A. Pletersek, "Integrierte Schaltung mit einem Magnetfeldsensor", Fakulteta za Elektrotehniko (Slovenia), European Patent, N° EP0655628, May 1995

[79] W. M. Frix, G. G. Karady, B. A. Venetz, "Comparison of Calibration Systems for Magnetic Field Measurement Equipment", IEEE Transactions on Power Delivery, Vol. 9, pp. 100-108, January 1994

[80] M. Demierre, "Improvements of CMOS Hall microsystems and application for absolute angular position measurements", PhD thesis N° 2844, Ecole Polytechnique Fédérale de Lausanne (EPFL), Lausanne, Switzerland, 2003

[81] J. Trontelj, "Optimization of Integrated Magnetic Sensor by Mixed Signal Processing", IEEE Instrumentation and Measurement Technology Conference, Vol. 1, pp. 299-302, May 1999

References 251

[82] LEM Current sensors, Product Catalog, LEM SA, Geneva, Switzerland, 2004

[83] P. L. C. Simon, P. H. S. de Vries, S. Middelhoek, "Autocalibration of Silicon Hall Devices", IEEE International Conference on Solid-State Sensors and Actuators, Vol. 2, pp. 237-240, June 1995

[84] J. Petr, H. Leinhard, "Compensating circuit for a magnetic field sensor", Landis & Gyr AG (Switzerland), United States of America Patent, N° US4752733, June 1988

[85] M. Demierre, S. Pesenti, R. S. Popovic, "Self calibration of a CMOS twin Hall microsystem using an integrated coil", Proceedings of the 16th European Conference on Solid-State Transducers (Eurosensors), pp. 573-574, September 2002

[86] P. Hazard, A. Boulahtit, "Dispositif de mesure de la sensibilité d'un capteur de champ magnétique à effet Hall", Schneider Electric Industries SA (France), French Patent, N° FR2796725, January 2001

[87] B. Lequesne, T. Schroeder, "High-Accuracy Magnetic Position Encoder Concept", IEEE Transactions on Industry Applications, Vol. 35, pp. 568-576, May-June 1999

[88] V. Frick, L. Hebrard, P. Poure, F. Braun, "CMOS Microsystem Front-End for MicroTesla Resolution Magnetic Field Measurement", IEEE International Conference on Electronics, Circuits and Systems, Vol. 1, pp. 129-132, September 2001

[89] Z. B. Randjelovic, M. Kayal, R. Popovic, H. Blanchard, "Highly Sensitive Hall Magnetic Sensor Microsystem in CMOS Technology", IEEE Journal of Solid-State Circuits, Vol. 37, pp. 151-159, February 2002

[90] V. Frick, P. Poure, L. Hebrard, F. Anstotz, F. Braun, "Design and Prototyping of a CMOS Standard Contactless Current Measurement Macrocell for Integrated Microsystems in Power Control Applications", IEEE International Conference on Electronics, Circuits and Systems, Vol. 2, pp. 645-648, September 2002

[91] M. Pastre, M. Kayal, H. Blanchard, "Continuously Calibrated Magnetic Field Sensor", LEM SA (Switzerland), European Patent, Application N° 04405584.6, September 2004

[92] M. Pastre, M. Kayal, H. Blanchard, "A Hall Sensor Analog Front End for Current Measurement with Continuous Gain Calibration", IEEE International Solid-State Circuits Conference, pp. 242-243, 596, February 2005

[93] D. Draxelmayr, R. Borgschulze, "A Self-Calibrating Hall Sensor IC with Direction Detection", IEEE Journal of Solid-State Circuits, Vol. 38, pp. 1207-1212, July 2003

[94] M. Motz, D. Draxelmayr, T. Werth, B. Forster, "A chopped Hall sensor with programmable "true power-on" function", IEEE European Solid-State Circuits Conference, pp. 443-446, September 2004

[95] C. Menolfi, Q. Huang, "A Fully Integrated, Untrimmed CMOS Instrumentation Amplifier with Submicrovolt Offset", IEEE Journal of Solid-State Circuits, Vol. 34, pp. 415-420, March 1999

[96] L. S. Milor, "A Tutorial Introduction to Research on Analog and Mixed-Signal Circuit Testing", IEEE Transactions on Circuits and Systems, Part II: Analog and Digital Signal Processing, Vol. 45, pp. 1389-1407, October 1998

[97] J. Frounchi, M. Demierre, Z. Randjelovic, R. S. Popovic, "Integrated Hall Sensor Array Microsystem", IEEE International Solid-State Circuits Conference, pp. 248-249, 452, February 2001

[98] M. Kayal, F. Burger, R. S. Popovic, "Magnetic Angular Encoder Using an Offset Compensation Technique", IEEE Sensors Journal, Accepted for future publication, 2004

[99] M. J. Hasler, M. Saghafi, A. Kaelin, "Elimination of Parasitic Capacitances in Switched-Capacitor Circuits by Circuit Transformations", IEEE Transactions on Circuits and Systems, Vol. 32, pp. 467-475, May 1985

[100] R. B. Datar, A. S. Sedra, "Exact Design of Strays-Insensitive Switched-Capacitor Ladder Filters", IEEE Transactions on Circuits and Systems, Vol. 30, pp. 888-898, December 1983

[101] K. Martin, A. S. Sedra, "Correction to "Effects of the Op Amp Finite Gain and Bandwidth on the Performance of Switched-Capacitor Filters"", IEEE Transactions on Circuits and Systems, Vol. 29, p. 198, March 1982

[102] H. Yoshizawa, Y. Huang, P. F. Ferguson, G. C. Temes, "MOSFET-Only Switched-Capacitor Circuits in Digital CMOS Technology", IEEE Journal of Solid-State Circuits, Vol. 34, pp. 734-747, June 1999

[103] G. W. Roberts, W. M. Snelgrove, A. S. Sedra, "Switched-Capacitor Realization of Nth-Order Transfer Function Using a Single Multiplexed Op-Amp", IEEE Transactions on Circuits and Systems, Vol. 34, pp. 140-148, February 1987

[104] G. Laimer, J. W. Kolar, "Wide Bandwidth Low Complexity Isolated Current Sensor to be Employed in a 10kW/500kHz Three-Phase Unity Power Factor PWM Rectifier System", IEEE Power Electronics Specialists Conference, Vol. 3, pp. 1065-1070, June 2002

[105] S. M. Mallya, J. H. Nevin, "Design Procedure for a Fully Differential Folded-Cascode CMOS Operational Amplifier", IEEE Journal of Solid-State Circuits, Vol. 24, pp. 1737-1740, December 1989

[106] J. Grilo, E. MacRobbie, R. Halim, G. Temes, "A 1.8V 94dB Dynamic Range Modulator for Voice Applications", IEEE International Solid-State Circuits Conference, pp. 230-231, 451, February 1996

[107] Y. Huang, G. C. Temes, H. Yoshizawa, "A High-Linearity Low-Voltage All-MOSFET Delta-Sigma Modulator", IEEE Custom Integrated Circuits Conference, pp. 293-296, May 1997

References

[108] G. W. Roberts, D. G. Nairn, A. S. Sedra, "On the Implementation of Fully Differential Switched-Capacitor Ladder Filters", IEEE Transactions on Circuits and Systems, Vol. 33, pp. 452-455, April 1986

[109] A. Chatterjee, N. Nagi, "Design for Testability and Built-In Self-Test of Mixed-Signal Circuits: A Tutorial", IEEE International Conference on VLSI Design, pp. 388-392, January 1997

Index

Numerics
1T DRAM 139

A
ADC
    Delta-Sigma 189
Algorithm
    1T DRAM reference
        adjustment 145, 147
    2-pass simulation 125, 134
    DAC calibration 81
    Radix conversion 84
    Reverse successive
        approximations 30
    Successive approximations 26
Amplifier 93, 201
    DDA 202
Autozero 11

B
Bias
    Current 80, 154

C
Calibration
    Continuous-time 115
    Gain 179
    Hall sensors 160, 163, 173
    M/2$^+$M 78
    Offset 93, 183
    Sensitivity 163, 173
    SOI 1T DRAM 138
    Table 86
Cascode 122, 207
    Regulated 88
Chopper stabilization 7
    in Hall sensors 157
Classification 21
Coil 162
Combined modulation 175
Compensation current technique 96
    Compensation node 105
    DAC resolution 113
    Decision filtering 114
    Detection configuration 97
    Detection node 100
    Simulation 124
Compensation node 105
Complementary M/2M ladder 49
Component arrays 35
    Sizing 36
Continuous-time compensation 115
Correlated double sampling 18
Current collectors 67, 87
Current comparator 89
Current division 41
Current sources 38
Current-mirror DAC 39

D
DAC resolution 113
DDA 202
Delta-Sigma ADC 189, 221

Demodulation
    Scheme   176, 227
Demodulator   208
    External signal   213
    Offset   220
    Reference   216
Detection configuration   97
Detection node   100
DRAM   138

E
External interferences   226
External signal demodulator   213

F
Filtering
    Noise   184

G
Gain compensation   179
Glitch   20

H
Hall
    Effect   152
    Sensor array   199
    Sensor microsystems   171
    Sensor models   155
    Sensors   151, 153
    Sensors calibration   160

I
Integrator
    Switched-capacitor   209
Interferences
    External   226
    Rejection   193

L
Ladder
    Complementary M/2M   49
    Current collectors   67
    Layout   72
    $M/2^+M$ 62
    M/2.5M   64
    M/2M   48
    M/3M   62
    R/2R   40
    R/xR   51
Layout
    $M/2^+M$ 72
Linear current division   41

M
$M/2^+M$ ladders   62
    Calibration   78
    Selection   66
M/2.5M ladders   64
M/2M ladders   48
    Complementary   49
    Trimming   51
M/3M ladders   62
Matching   5
    Parameters   6
    Rules   6
Measurements
    Hall microsystem   233
    $M/2^+M$ 73
    SOI 1T DRAM   147
Memory   138
Microsystem   171
Miller operational amplifier   93
Modulation
    Combined   175
    Scheme   175, 227
Monte Carlo   136

N
Noise   14, 114, 184

O
Offset compensation   183
Offset demodulator   220
Operational amplifier   93, 207

P
Ping-pong   18
Preamplifier   201
PSpice   134

R
R/2R ladders   40

Index 257

R/xR ladders   51
    Radix   60
    Sizing   57
    Terminator calculation   54
    Terminator implementation   55
    Terminator sizing   58
    Working condition   53
Radix   60
    Conversion algorithm   85
Reference demodulator   216
    Offset compensation   224
Regulated cascode   88
Rejection of interferences   193
Resolution   113

S

Scheme
    Demodulation   177, 228
    Modulation   176, 228
Sense amplifier   142
Sensitivity calibration   163, 173
Sensitivity drift   161
Simulation
    Digital compensation   124
    Monte Carlo   136
    Multiple digital compensation   133
SOI 1T DRAM   138
Spinning current   157, 173
Sub-binary radix DACs   31
    and successive approximations   31
    Characteristics   32
    Resolution   34
    Tolerance to radix variations   34
Successive approximations   24
    Algorithm   26
    and sub-binary radix DACs   31
    Complexity   31
    Reverse algorithm   30
    Working condition   28
Switch   10
Switch box   158
Switched-capacitor   209

T

Temperature   90, 161
Terminator
    Implementation   55
    Sizing   58

Tolerance to radix variations   34
Transmission gate   10

U

Up/down DAC   117

W

Working condition   28, 53